Arsenic Monitoring, Removal and Remediation

Edited by Margarita Stoytcheva and Roumen Zlatev

Published in London, United Kingdom

IntechOpen

Supporting open minds since 2005

Arsenic Monitoring, Removal and Remediation
http://dx.doi.org/10.5772/intechopen.95693
Edited by Margarita Stoytcheva and Roumen Zlatev

Contributors
Bartlomiej K. Konstanty Bancewicz, Mohammed Zia Uddin Kamal, Md. Yunus Miah, Tamar Varazi, Gia Khatisashvili, Maritsa Kurashvili, Marine Pruidze, Kakha Didebulidze, Tinatin Butkhuzi, Nina Asatiani, Tamar Kartvelishvili, Nelly Sapojnikova, Elina Bakradze, Evgeni Bunin, Kaman Singh, Ashok Kumar, Utkarsh Dixit, Rayees Ahmad Bhat, Satya Prakash Gupta, Margarita Stoytcheva, Roumen Zlatev

First published in London, United Kingdom, 2022 by IntechOpen
IntechOpen is the global imprint of INTECHOPEN LIMITED, registered in England and Wales, registration number: 11086078, 5 Princes Gate Court, London, SW7 2QJ, United Kingdom
Printed in Croatia

British Library Cataloguing-in-Publication Data
A catalogue record for this book is available from the British Library

Additional hard and PDF copies can be obtained from orders@intechopen.com

Arsenic Monitoring, Removal and Remediation
Edited by Margarita Stoytcheva and Roumen Zlatev
p. cm.
Print ISBN 978-1-83969-665-7
Online ISBN 978-1-83969-666-4
eBook (PDF) ISBN 978-1-83969-667-1

For EU product safety concerns:
IN TECH d.o.o., Prolaz Marije Krucifikse Kozulić 3, 51000 Rijeka, Croatia, info@intechopen.com or visit our website at intechopen.com.

Meet the editors

Prof. Margarita Stoytcheva graduated from the University of Chemical Technology and Metallurgy of Sofia, Bulgaria. She has a Ph.D. and DSc in Chemistry and Technical Sciences. She has been a researcher and teacher at several universities in Bulgaria, Algeria, and France. From 2006 to the present, she has participated in activities of scientific research, technological development, and teaching at the Institute of Engineering, University of Baja California, Mexicali, Mexico, as a full-time researcher. Since 2008 she has been a member of the National System of Researchers of Mexico, and since 2011 she has been a regular member of the Mexican Academy of Sciences. Her interests and areas of research are electroanalytical chemistry and biotechnology.

Dr. Zlatev obtained his master's degree from the University of Chemical Technology and Metallurgy of Sofia, Bulgaria, and his Ph.D. from the Grenoble Institute of Technology, France. After his work as a researcher at the Bulgarian Academy of Sciences, Dr. Zlatev began working as a senior researcher and laboratory head at the Institute of Engineering, University of Baja California, Mexicali, Mexico. He is a regular member of the Mexican Academy of Sciences and the Mexican National System of Researchers. Dr. Zlatev has authored more than 110 publications in prestigious scientific journals and holds 10 patents in analytical and electroanalytical chemistry, spectroscopy, corrosion, and analytical instrumentation.

Contents

Preface

Arsenic Monitoring, Removal and Remediation is organized into two sections containing five chapters.

The two chapters in the first section comment on the principal methods for determining arsenic levels in environmental samples, such as atomic absorption spectrometry, inductively coupled plasma mass spectrometry, inductively coupled plasma-atomic emission spectrometry, X-ray fluorescence, molecule adsorption spectrophotometry, and others, with emphasis on sample pretreatment, extraction, separation, and method validation techniques for arsenic speciation analysis. Special attention is paid to the electrochemical methods for arsenic quantification as an alternative to the aforementioned techniques and to the potential of the differential alternative pulses voltammetry for arsenic determination in the presence of interferences.

The three chapters in the second section highlight the benefits of using adsorption for arsenic species removal and suggest remedial approaches against arsenic pollution. The strategies discussed include traditional arsenic treatment technologies such as oxidation, coagulation-flocculation, adsorption, ion exchange, and membrane filtration, along with more sustainable arsenic remediation methods such as bioremediation.

The book offers solutions to problems associated with arsenic speciation, determination, and remediation. It launches a call to action to avoid arsenic pollution.

Margarita Stoytcheva and Roumen Zlatev
Universidad Autónoma de Baja California,
Mexicali, Mexico

Section 1

Arsenic Speciation Techniques and Determination

Chapter 1

Introductory Chapter: Advanced Electrochemical Technique for Arsenic Determination in Complex Samples

Margarita Stoytcheva and Roumen Zlatev

1. Introduction

Arsenic contamination is a serious environmental and health threat worldwide. Pollution is due to various natural and artificial processes as degradation of industrial waste, pesticide leaching, rock and mineral erosion with the contribution of acid rains, magmatic processes, industrial practices such as mining, metal smelting, coal-fired power plants emissions, etc. Arsenic determination in environmental samples is carried out applying numerous methods: atomic absorption spectrometry, inductively coupled plasma mass spectrometry, neutron activation analysis, inductively coupled plasma-atomic emission spectrometry, X-ray fluorescence, molecule adsorption spectrophotometry, etc. [1, 2]. Unfortunately, these methods are complicated and require sophisticated and expensive equipment. In addition, most of them are used for total arsenic content determination, i.e. they do not allow the quantification of the individual arsenic species, whose distribution in the environmental and health effects are different. This work describes some of the properties of arsenic and the electrochemical methods applied to its determination, as an alternative to the aforementioned techniques, with special emphasis on the differential alternative pulses voltammetry (DAPV) as an advanced electrochemical method for arsenic determination in complex samples.

2. Arsenic and its properties

Arsenic is a chemical element, which exists in various inorganic and organic forms.

Inorganic arsenic species mainly include As(V): H_3AsO_4, $H_2AsO_4^-$, $HAsO_4^{2-}$, and AsO_4^{3-}. H_3AsO_4 is the most commonly found in the environment. It is the predominant arsenic species in natural waters, soils, and drinking water. It is also present in plants, algae, and aquatic animals. Arsenates have a high ionization capacity. The molecule, by losing the hydrogen ion by dissociation, remains negatively charged, forming several anions. The As(III) species: H_3AsO_3, $H_2AsO_3^-$, $HAsO_3^{2-}$, and AsO_3^{3-} are considered as the most toxic. The oxidation state of arsenic depends on the conditions, such as the redox potential and pH. Under oxidizing conditions As(V) predominates over As(III) in the form of $H_2AsO_4^-$ (pH < 6.9) and in the form of $HAsO_4^{2-}$ (at higher pH). At pH < 2 (extremely acidic) the dominant species is H_3AsO_4. At pH > 12 (extreme basicity) the dominant species is AsO_4^{3-}. Under reducing conditions and pH < 9.2 the neutral specie H_3AsO_3 prevails. In surface waters As(V) predominates over As(III), while in groundwater both may be present.

Organic arsenic species are widely distributed in the atmosphere, aquatic systems, soils, sediments, and biological tissues. They are involved in biologically mediated methylation, which occurs in terrestrial and marine organisms and which converts inorganic arsenic into nontoxic methylated compounds or in methylated compounds of moderate toxicity. Organic forms of arsenic usually occur in lower concentrations than inorganic species, although their ratio may increase as a result of methylation reactions catalyzed by microbial activity (bacteria, algae). The predominant organic forms are: dimethylarsinic acid and monomethylarsonic acid, where arsenic is present in both cases as As(V).

The toxicity of arsenic species decreases in the following order: AsH_3 > As(III) species > As(V) species > organic arsenic species. The lethal dose of As(III) for adults is 1–4 mg/kg body weight. For the other arsenic compounds, the lethal dose varies between 1.5 and 500 mg/kg. The human body is particularly protected by the methylation of arsenic, producing less toxic and more extractable metabolites. The World Health Organization advises that the maximum allowed arsenic concentration in drinking water is 10 $\mu g\ L^{-1}$.

3. Electrochemical methods for arsenic determination

Electrochemical methods provide accurate measurements of As concentration in the ppb range, using simple and cheap equipment. The most commonly used techniques are square wave voltammetry (SWV) and differential pulse voltammetry (DPV) including their stripping mode [3–5]. Arsenic detection is performed applying various bare (Pt, Au, carbon) and chemically modified electrodes [6, 7]. Unfortunately, in contrast to the mercury electrode, the application of all types of solid electrodes is limited by the so-called "memory effect". As an alternative, the hanging mercury drop electrode (HMDE) with a renewed mercury surface or mercury film electrodes (MFE) can be used. However, their application is limited by the toxicity of mercury.

The measurement of As(III) concentration applying voltammetry can be complicated due to the interference of other metals present in the sample such as: Pb(II), Cu(II), Sb(III), Ag(I), Se(IV), Bi(III), Hg(II), Fe(II), Tl(I). As their peak potentials are close to the peak potential of As(III), peaks overlapping is observed, which impedes the precise As(III) peak height evaluation and hence – the precise As(III) concentration determination.

Recently, Zlatev et al. [8] developed a high-resolution voltammetric technique: differential alternative pulses voltammetry (DAPV) allowing the simultaneous

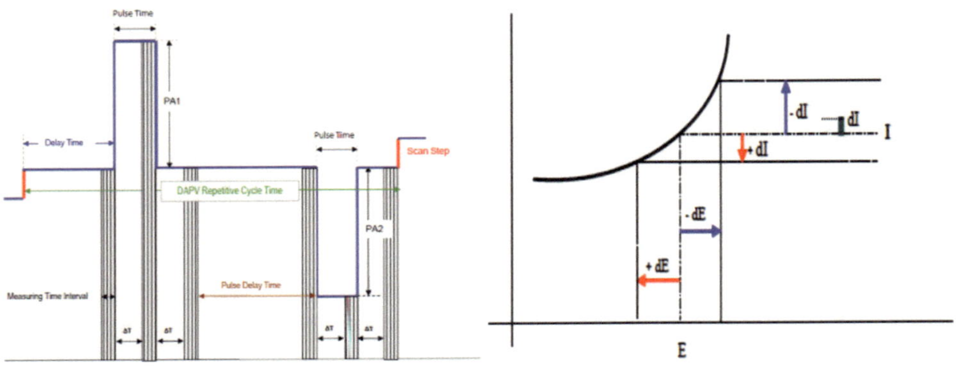

Figure 1.
DAPV potential-time waveform (left) and the corresponding current response (right).

quantification of species having very close peak potentials. DAPV potential-time waveform and the corresponding current responses are shown in **Figure 1**. This technique provides the simplicity and the high sensitivity of the DPV combined with the high resolution of the second-order voltammetric techniques [9–13]. The high resolution of the DAPV is due to the registered curve shaped as the first derivative of a peak consisting of a cathodic and an anodic peak with a small half-width for any of the species. In the case of very close peak potentials, one of the species is determined by its cathodic peak, while the other by its anodic peak remained on the voltammogram after the overlapping [8, 9].

The anodic dI_{p+} and the cathodic dI_{p-} current responses can be expressed by the following equations for reversible electrochemical reactions:

$$dI_{p_-} = \frac{n^2F^2}{RT}AC(-dE)\sqrt{\frac{D}{\pi t}}\frac{P_-}{\left(1+P_-\right)^2} \qquad P_- = \exp\left[\left(E - E_{\frac{1}{2}} - \frac{dE}{2}\right)\frac{nF}{RT}\right]$$

$$dI_{p_+} = \frac{n^2F^2}{RT}ACdE\sqrt{\frac{D}{\pi t}}\frac{P_+}{(1+P_+)^2} \qquad P_+ = \exp\left[\left(E - E_{\frac{1}{2}} + \frac{dE}{2}\right)\frac{nF}{RT}\right]$$

where R is the gas constant, T is the absolute temperature, n is the number of transferred electrons, F is the Faraday constant, A is the electrode area, C is the concentration, dE is the pulse amplitude, D is the diffusion coefficient, t is the current measurement delay time, E is the electrode potential and $E_{1/2}$ is the half-wave potential.

DPV curve of a water sample containing 1 μmol L^{-1} As(III) and 1 μmol L^{-1} Pb(II) in 1 mol L^{-1} HCl supporting electrolyte is shown in **Figure 2** left. The DPV peaks of the two species are completely overlapped forming a common peak even in concentration ratio 1:1, which makes its distinguishing and determination impossible. Separate peaks however are registered on the DAPV curve at the same conditions as seen in **Figure 2** right. The two species can be distinguished even in concentration ratio As(III)/Pb(II) as high as 4 to 1 as previously reported by Zlatev et al. [14].

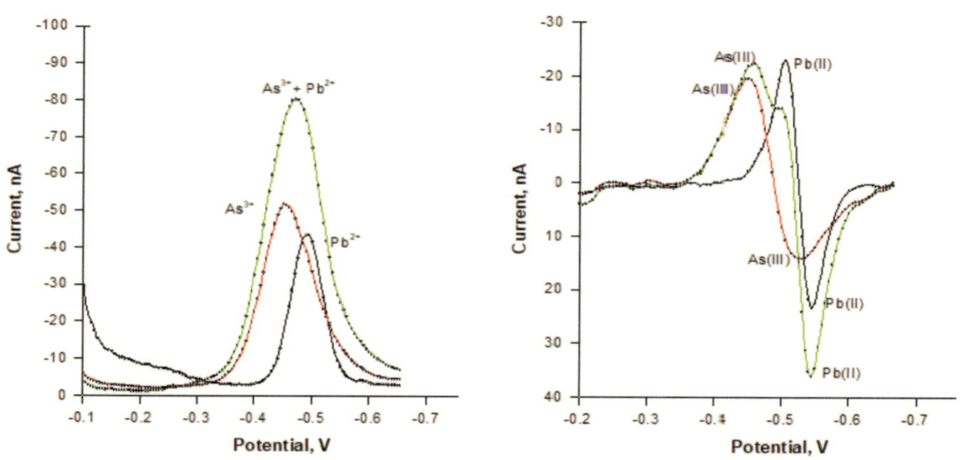

Figure 2.
DPV (left) and DAPV (right) voltammograms of 1 μmol L^{-1} As(III) and 1 μmol L^{-1} Pb(II) at HMDE. Pulse amplitude = 10 mV, scan rate = 10 mV s^{-1}, scan step = 5 mV.

4. Conclusion

The importance of measuring the arsenic concentration is due to the great environmental pollution with this toxic and carcinogenic element. The methods currently used for this purpose (atomic absorption spectrometry, inductively coupled plasma, chromatography, etc.) require complicated and expensive equipment. In this context, electrochemical methods represent an attractive alternative. Special attention has to be paid to the advanced technique differential alternative pulses voltammetry allowing the precise arsenic determination in complex samples despite the interference of the species with close peak potentials.

Author details

Margarita Stoytcheva* and Roumen Zlatev
Instituto de Ingeniería, Universidad Autónoma de Baja California, Mexicali, Mexico

*Address all correspondence to: margarita.stoytcheva@uabc.edu.mx

References

[1] Stoytcheva M, Sharkova V, Panayotova M. Electrochemical approach in studying the inhibition of acetylcholinesterase by arsenate (III)—Analytical characterisation and application for arsenic determination. Analytica Chimica Acta. 1998;**364**:195

[2] Luong JHT, Majid E, Male KB. Analytical tools for monitoring arsenic in the environment. The Open Analytical Chemistry Journal. 2007;**1**:7

[3] Antonova S, Zakharova E. Inorganic arsenic speciation by electroanalysis. From laboratory to field conditions: A mini-review. Electrochemistry Communications. 2016;**70**:33

[4] Forsberg G, O'Laughlin JW, Megargle RG, Koirtyihann SR. Determination of arsenic by anodic stripping voltammetry and differential pulse anodic stripping voltammetry. Analytical Chemistry. 1975;**47**:1586

[5] Mays DE, Hussam A. Voltammetric methods for determination and speciation of inorganic arsenic in the environment—A review. Analytica Chimica Acta. 2009;**646**:6

[6] Cox JA, Rutkowska IA, Pawel J, Kulesza PJ. Critical review—Electrocatalytic sensors for arsenic oxo species. Journal of the Electrochemical Society. 2020;**167**:037565

[7] Kempahanumakkagari S, Deep A, Kim KH, Kumar Kailasa S, Yoon HO. Nanomaterial-based electrochemical sensors for arsenic—A review. Biosensors & Bioelectronics. 2017; **95**:106

[8] Zlatev R, Stoytcheva M, Valdez B, Magnin J-P, Ozil P. Simultaneous determination of species by differential alternative pulses voltammetry. Electrochemistry Communications. 2006;**8**: 1699

[9] Zlatev R, Stoytcheva M, Valdez B. Application of anodic stripping differential alternative pulses voltammetry for simultaneous species quantification. Electroanalysis. 2018;**30**:1902

[10] Agarwal HP, Saxena M. Faradaic rectification polarographic studies of reduction of zinc ions in indifferent electrolytes. Indian Journal of Chemistry A. 1978;**16**:754

[11] Barker GC, Gardner AW, Williams MJ. A multi-mode polarography. Journal of Electroanalytical Chemistry. 1973;**42**:A21

[12] Saur D. Differential faradaic rectification polarofraphy. Fresenius' Zeitschrift fuer Analytische Chemie. 1979;**298**:128

[13] Neeb RZ. Oberwellen-Wechselstrompolarographie. Fresenius' Zeitschrift fuer Analytische Chemie. 1962;**188**:401

[14] Zlatev R, Stoytcheva M, Valdez B. As(III) determination in the presence of Pb(II) by differential alternative pulses voltammetry. Electroanalysis. 2010;**22**: 1671

Chapter 2

Arsenic Speciation Techniques in Soil Water and Plant: An Overview

Mohammed Zia Uddin Kamal and Md. Yunus Miah

Abstract

There are more than 100 different arsenic with different characteristics in the soil-water-plant ecosystem. The identification and quantification of individual arsenic species is essential for understanding the distribution, environmental fate and behavior, metabolism and toxicity of arsenic. Due to the hazardous nature of arsenic, people have a high interest in the measurement of arsenic species. The reaction of the formation of arsenic speciation in the soil-water-plant environment is briefly studied. There is little information on methods used to quantify arsenic forms and species in contaminated soil, water and plant. The purpose of this article is to understand the available sample pretreatment, extraction, separation, detection and method validation techniques for arsenic speciation analysis of arsenic species in soil, water and plant. The performances of various sample preparation and extraction processes, as well as effective separation techniques, that contribute greatly to excellent sensitivity and selectivity in arsenic speciation when coupling with suitable detection mode, and method validity are discussed. The outlines of arsenic speciation techniques are discussed in view of the importance to the completeness and accuracy of analytical data in the soil-water-plant samples. To develop cheap, fast, sensitive, and reproducible techniques with low detection limits, still needed to confine research on arsenic speciation present in environmental matrices.

Keywords: Arsenic speciation, extraction, separation, detection, techniques, soil water and plant

1. Introduction

Arsenic (As) is a geogenic toxic metalloid found ubiquitously in environmental systems such as lithosphere (earth crusts, soil, rock, and sediment), hydrosphere (surface water, aquifers, deep wells, and oceans), atmosphere and biosphere (food chain and ecosystems) [1]. Arsenic is considered as the 12th most abundant elements in the earth's crust. Elevated arsenic having been introduced in the ecosystem either by natural routes involve in weathering and other biogeochemical processes or via anthropogenic activities, including mining, and smelting, excessive agricultural utilization of As-based fertilizers and pesticides and irrigation with As-laden groundwater [2–4]. This problem becomes serious concern because once arsenic is released in the soil and water resources, it is bioaccumulated by the terrestrial and aquatic biota, and subsequently enters in the human or animal food chain [5, 6]. In highly arsenic contaminated ($\geq$0.01 mg L^{-1}) area, the migration of arsenic from soil to water and plant is a serious problem, becoming a major threat to sustainable agriculture practices and food security [7, 8]. Empirical data shows that the concentration of arsenic in contaminated soils lies between 10 mg kg^{-1} and as high as 30,000 mg/kg [9]. In addition, the reported concentrations of arsenic in all natural waters is between <0.5 μg L^{-1} and more than 5000 μg L^{-1}, although maximum permissible contaminant total As limits in drinking water by WHO is 10 μg L^{-1} [1, 10]. Moreover, considering toxicity, the Joint Food and Agriculture Organization and the World Health Organization (FAO/ WHO) Expert Committee on Food Additives proposed that the maximum inorganic arsenic content in food such as polished rice is 0.2 mg kg^{-1} [11–13]. Thus, exposure to arsenic (As) in soil-water-plant becomes global public health and the environment concern due to the wide distribution in ecosystem and its close association with numerous adverse effects.

There are more than 100 different arsenic compounds in the soil-water-plant ecosystem [14, 15]. It is well known that the toxicity, bioavailability, physiological and metabolic processes and mobility of arsenic vary greatly depending on the chemical species and oxidative states rather than its total content [16, 17]. Arsenic (As) speciation analysis may specify not only the determination of total As contents but also considering its specific ionic forms in the aqueous solution and the sequential extracted As related to various mineral phases [18]. According to the IUPAC recommendations, "speciation of an element" is defined as "the distribution of an element amongst defined chemical species in a system" rather than fractionation. While speciation analysis is defined as "analytical activity of measuring the quantities of one or more individual chemical species in a sample" [19].

Arsenic exists multiple oxidation states (+III, +V, 0, –III) and various inorganic and organic chemical species. In environmental assessment, it is far from enough to know the total arsenic content in actual samples, because the toxicity of As element is predestined by distinct arsenical species [20]. Generally speaking, trivalent arsenic compounds are usually more toxic than pentavalent arsenic compounds [4] and inorganic species are more toxic than the organic ones. Again, trivalent organic arsenicals can be more toxic than trivalent inorganic arsenicals [21]. The United States Environmental Protection Agency (USEPA) priority pollutants list represents inorganic As is the first category of toxins [22], classified as Group I carcinogens based on human epidemiological data. In addition, the organic species toxicity usually decreases with the increase of methylation. For example, monomethylarsonic acid (MMA) and dimethylarsinic acid (DMA) are less toxic, arsenobetaine (AsB), arsenocholine (AsC) and other arsenosugars are even considered non- toxic [21]. However, in certain environments, such as in aquatic biomass, AsB can be converted into toxic inorganic arsenicals and enter the food chain [23, 24]. Depending on the source, metals or metalloids may enter the environment, where they may be transformed into another compound. Therefore, As speciation is essential for understanding its distribution, transformation in the environment, toxicity, metabolism, bioavailability and health effects in the natural system [15].

There is a huge difference between the toxicity and distribution of the arsenic species observed in the environment, which heights the importance of detecting and quantifying a single compound. Recently, various techniques have been developed to figure out arsenic species in environmental and biological samples, including soil, water and plants. The establishment of the new arsenic speciation analysis program not only improves our understanding of arsenic biogeochemistry, toxicity and metabolism but also provides a lot of information about exposure biomarkers and arsenic cycling in the natural environment. However, it is still a challenge to completely isolate the target arsenic compound from background interference [25]. Therefore, a quick and simple method is needed to analyze the arsenic species in different matrices. In addition, optimizing the extraction of target arsenic is also crucial for accurate quantification [26]. Determining the exact species of arsenic in biological and environmental samples helps to more accurately assess the environmental impact and health risks caused by arsenic exposure. Appropriate sample pretreatment techniques are necessary to reduce the influence of matrix, to enrich the target species and/or separation of As species for accurate identification. The newly developed As speciation protocols must achieve suitable detection mode, excellent selectivity and sensitivity in various environmental and biological species. Various non-chromatographic and chromatographic methods are involved in the selective separation of the arsenic species.

To date, several study on overall arsenic speciation analysis have been done [15, 27, 28]. Nevertheless, there is still a big knowledge gap in the speciation of arsenic. This overview includes arsenic speciation analysis, species detection systems, key extraction/separation techniques and mechanisms used in the accuracy assessment of speciation methods, and focuses on important strategies for specific arsenic speciation. This study will provide sentinels on comprehensively discuss in the latest developments in arsenic speciation analysis and challenges for further research.

2. Reactions of arsenic speciation on environment

Arsenic is introduced into the environment either naturally or anthropogenically; once released, it cannot be degraded or destroyed. The environmental transformation of arsenic depends on the availability of arsenic in the geological source, as well as their oxidation state, speciation and other environmental factors [29, 30]. There are different forms of arsenic containing mineral in the Earth's crust. For example, 60% are in arsenate form, 20% are in the form of sulfides and sulfonates, and the remaining 20% are in the form of arsenites, arsenides, silicates, oxides, and elemental As [31]. In the soil and water environments, As can exist in four different oxidation states (As^{3+}, As^{5+}, As^{0} and As^{3-}), in inorganic as well as in organic forms [4]. The most widespread arsenic species detected in the environment and biological systems are shown in **Table 1**.

In natural environment, inorganic arsenic contains two oxygen anions, arsenite As (III) and arsenate As (V) but there are many organic arsenic compounds including monomethyl arsonic acid (MMA) and dimethyl arsinic acid (DMA) is the most common. According to intake and mobility, the toxicity of arsenic compounds decreases in the following sequential order: arsines > inorganic arsenites > organic trivalent compounds (arsenoxides), inorganic arsenates > organic pentavalent compounds > arsonium compounds > elemental arsenic. Arsenobetaine and arsenocholine are considered nontoxic [4]. At the same time, arsenic species exhibit various reaction behaviors and metabolic transformations in soil-water and plant ecosystems. For arsenic risk assessment of environmental samples and detection of appropriate speciation analysis, it is necessary to understand the main forms and metabolic transformations of arsenic compounds.

2.1 Arsenic speciation in soil

The various species or chemical forms of As in soil include- free ionic forms, precipitated as solids, adsorbed on soil organic or inorganic constituents, exchangeable, and structural constituent of primary and secondary minerals [32, 33]. There are both inorganic and organic forms (species) of arsenic in the soil. The most common inorganic species are arsenate (AsV) and arsenite (AsIII), while the most common organic species are monomethylarsonic acid (MMA) and dimethylarsinic acid (DMA). The order of toxicity of arsenic species is AsIII > AsV > MMA > DMA [34]. In general, minor amount of naturally occurring arsenic in soil exists as a form of amorphous iron and aluminum oxides.

Name	Abbreviation	Chemical structure
Inorganic form		
Arsenite (Arsenous acid)	As (III)	H_3AsO_3, H_2AsO^{3-}, $HAsO_3^{2-}$, AsO_3^{3-}
Arsenate (arsenic acid)	As (V)	H_3AsO_4, $H_2AsO_4^-$, $HAsO_4^{2-}$, AsO_4^{3-}
Organic form		
Monomethylarsenic acid	MMA	$CH_3AsO(OH)_2$
Monomethylarsonic acid	MMA(V)	$CH_3AsO(OH)_2$
Monomethylarsonous acid	MMA(III)	$CH_3As(OH)_2$
Dimethylarsinic acid	DMA (V)	$(CH_3)_2AsO(OH)$
Dimethylarsinous acid	DMA (III)	$(CH_3)_2AsOH$
Dimethylarsinoyl ethanol	DMAE	$(CH_3)_2AsOCH_2\ CH_2OH$
Trimethylarsine oxide	TMAO	$(CH_3)_3AsO$
Trimethylarsoniopropionate	TMAP	$(CH_3)_3As^+CH_2CH_2COO^-$
Tetramethylarsonium ion	TETRA, TMA	$(CH_3)_4As^+$
Arsenobetaine	AsB	$(CH_3)_3As^+CH_2COOH$
Arsenobetaine 2	AsB-2	$(CH_3)_3As^+CH_2CH_2COO^-$
Arsenocholine	AsC	$(CH_3)_3As^+CH_2CH_2OH$
Trimethylarsine	TMA (III)	$(CH_3)_3As$
Arsines	AsH_3, $MeAsH_2$, Me_2AsH	$(CH_3)_xAsH_{3-x}$ (x = 0–3)
Ethylmethylarsines	Et_xAsMe_{3-x}	$(CH_3CH_2)_xAs\ (CH_3)_{3-x}$ (x = 0–3)
Dimethylarsinyolacetic acid	DMAA	$(CH_3)_2AsOCH_2\ COOH$
Phenylarsine oxide	PAO	C_6H_5AsO
Phenylarsonic acid	PAA	$C_6H_5AsO(OH)_2$
Arsenosugars $C_7H_{14}AsO_3CH_2CH(OH)CH_2R$		
Arsenosugar 1 (glycerol sugar)	—	R = OH
Arsenosugar 2 (phosphate sugar)	—	R = $OP(O)(O^-)OCH_2CH(OH)CH_2OH$
Arsenosugar 3 (sulphonate sugar)	—	R = SO_3^-
Arsenosugar 4 (sulphate sugar)	—	R = OSO_3^-

Table 1.
Arsenic species commonly identified in the environment and biological systems.

Arsenic can be transformed in the soils through various mechanisms, such as oxidation, reduction, adsorption, dissolution, precipitation, and volatilization. The inorganic species of As, As(III) and As(V), are present in different forms (e.g., fully protonated As acids or arsenous acid) [35]. The main and thermodynamically stable forms of As(V/III) in soil may include $H_2AsO_4^-$, $HAsO_4^{2-}$ and H_3AsO_3. The existence of different As forms in soil largely depends on the texture, organic matter, pH value and redox potential of the surrounding environment. Arsenic exists in aerobic soil (oxidized conditions) in the form of arsenate (As^V) and is rapidly adsorbed on clay minerals and Fe/Mn oxides/hydroxides [2]. However, in reducing soil environment such as paddy fields, the arsenite (AsIII) form of arsenic dominates and its solubility, mobility, and toxicity are about 60 times that of As(V) [2]. In addition, under anaerobic conditions in the presence of sulfides, arsenic may precipitate in the form of arsenic sulfide and release excess arsenic into the environment [36]. Anaerobic bacteria degrade into less toxic volatile forms, such as dimethyl arsenic acid (DMAA) and monomethyl arsenic acid (MMAA) [37]. The oxidation and reduction of arsenic species takes place biologically and chemically in soil and water [38]. In addition, higher concentrations of arsenic were observed in alluvial soils and organic soils, while lower concentrations of arsenic were found in sandy soils [39]. Clay played a leading role in arsenic fixation. Arsenate is mainly adsorbed on clay particles in soils with neutral pH. Soil pH plays a major role in the types of arsenic compounds present in the soil. Under acidic conditions, arsenic tends to form compounds with aluminum and iron ($AlAsO_4$, FeAsO4), whereas under alkaline conditions (limestone soils) Under acidic conditions, arsenic tends to form compounds with aluminum and iron.

Arsenate exists in the form of oxygen anions at neutral pH, while arsenite has a neutral charge at pH 7.0. It leads to the formation of $Ca_3(AsO_4)_2$, $Mn_3(AsO_4)_2$, $AlAsO_4$ and $FeAsO_4$ [40]. However, when the pH value is higher than 8.5, as the pH value increases, the adsorption capacity of As^V decreases, while the case of As^{III} is the opposite. But at pH around 4, the adsorption for As^V on FeOOH is maximum, whereas for As^{III} the optimum pH is 7–8.5 [41]. Arsenic is more soluble under high or low pH values. In reducing environment, as arsenate is reduced to arsenite, it binds less strongly to the hydroxide solids, which increase the bioavailability of arsenic [42]. On the contrary, due to the larger arsenic sorption affinity, organic matter has formed organo-arsenic complex and reduced the solubility of arsenic in the soil [43]. Naturally, arsenic can be released into the soil environment through the hydrolysis and oxidation process of primary sulfide mineral (i.e arsenopyrite) and absorbed by ferric hydroxide. Meanwhile, the retention and mobilization of arsenic in soil is largely controlled by Ferric hydroxide. Such as iron oxides has stronger arsenic adsorption capacity than manganese oxides. Moreover, Phosphate plays significant role in absorption of arsenic from contaminant soil. Williams et al. [44] reported that in iron oxides dominated acidic soil, approximately 60% of the adsorbed pentavalent arsenic and 70% of the trivalent arsenic were displaced by H_2PO_4. Soil microbial activities can affect the adsorption/desorption, solubility, bioavailability, mobility and soil–plant transfer of arsenic by changing the chemical speciation of As in soil [45, 46]. Due to the action of microorganisms or the past use of methylated arsenic compounds, dimethyl sulfoxide, or sodium salts of MMA and DMA as pesticides, methylated As species, (i.e MMA and DMA) might be accumulate in soil [45]. Microorganisms in soil may interconvert As species As^V and As^{III} and further transform into MMA and DMA.

2.2 Arsenic speciation in water

The presence of arsenic in water is either dissolved or in particulate form. Arsenic pollution in groundwater mainly occurs due to release of geothermal water, desorption and reductive dissolution of iron oxides and oxidation of sulfide minerals [37]. High levels of arsenic in groundwater have been observed in many countries, such as more than 3000 μg L^{-1} because As has been released geogenically either by oxidation of arsenopyrite, or by reductive dissolution of arsenic rich ferrous oxyhydroxides in reducing aquifer environment [47, 48]. The most common forms of arsenic in natural waters are arsenite and arsenate. However, the main species found in natural water are forms of inorganic arsenic, namely H_2AsO_4-, H_3AsO_3, $HAsO_4^-$ and $As_3O_4^{3-}$. The change in the arsenic solubility in sulfidic water promotes the formation of amorphous metal-sulfide complex thioaresenic compounds [49]. Various species of aquatic micro and higher organisms plays important role on biomethylation process of arsenic, which reduces As^{5+} to soluble As^{3+} species [37]. Arsenate is the main form of arsenic in seawater. Dimethyl arsenic acid (DMMA) and methylarsenic acid (MMA) are also present in small amounts in seawater [50]. Moreover, the determination of arsenic in saline waters bear much importance due to gaining knowledge on because salinity induced inorganic arsenic specifically arsenite transformation to arsine gas [51].

2.3 Arsenic speciation in plants

Arsenic is not essential elements for plants development, although very small amounts of As in plants may have a positive effect on plant species. The concentration of As in plants is usually less than 1.0 mg kg^{-1} dry weight (DW) [52]. The mechanism by which plants absorb arsenic varies depending on the chemical form of arsenic.

2.3.1 Transformation of inorganic arsenic species in plants

Plants absorb inorganic arsenic through two mechanisms. The first mechanism involves the use of a high-affinity PO_4 transporter through phosphate (PO_4) transport pathway [53, 54] which uptake As (V) from soil solution and subsequently to aerial parts of the plants [55]. While, the second route for plant roots to absorb arsenic is through the aquaporin channels, which uptake As (III) (silicic acid analog) and methylated As species (MMA and DMA) [56]. In rice root cells, As (III) uses generally Si transporter owing to its similarities with silicic acid. Once in plant cell, As (V) is reduced to As (III) with the help of As reductase, ACR2 [57]. The detoxification of As (III) is achieved by forming complexes with thiol- rich peptides [58]. The form of As in the phloem is considered to be very crucial for the redistribution of As in various tissues in the plant [59].

2.3.2 Organic arsenic species transformation in plant

The methylated arsenic species, such as MMA and DMA, contribute less to the total arsenic in the soil. The organic arsenic substances MMA and DMA have taken up by the intrinsic protein of Oncokin 26. Rabb et al. [60] showed that the absorption efficiency of inorganic As species (AsIII and AsV) in roots is much higher than that of methylated As species (DMA and MMA), but the translocation efficiency of inorganic species in plant stems is much lower than that of methylation As species. The decrease in the As complex formed with ligands (such as glutathione) in the root may be the reason for the better translocation of methylated As species [60]. In rice, As (III) is found to be the most abundant species, followed by DMA. As (V), MMA and two other unidentified As also have found in lower concentrations [61]. A speciation analysis revealed the As (V) as a predominant species in rice straw followed by As (III) and DMA [62]. Meanwhile, in rice grain, As (III) and DMA are the dominant species.

3. Soil, water and plant sample preparation and extraction of arsenic species

3.1 Soil, water and plant sample preparation for arsenic speciation

Sample preparation and storage procedures are considered to be a key prerequisite for maintaining the concentration and chemical structure of the original species in the sample during the analysis process to obtain accurate As speciation information. Impractical As speciation data may arise due to losses during sampling, unrepresentative samples [63], contamination, mutual conversion between species, inefficient extraction of the analyte, and the possibility of precipitation and wall effects from the sample container [64]. To obtain reliable arsenic speciation data in soil, water and plant samples, two main strategies should be considered. First, determine appropriate species preservation practices to keep the chemical species of interest unchanged throughout the analysis process through avoiding changes in oxidation state, changes caused by microbial activity, and losses caused by volatilization or adsorption. Secondly, the species can be quantitatively converted into appropriate derivatives for further separation, accumulation and quantification [65]. Microbial transformation of arsenical compounds in contaminated samples is observe through a change in valence (i.e.oxidation/reduction) or chemical form (i.e. solid, liquid and gas) and formed biomethylate arsenic and both volatile (e.g., methylarsines) and nonvolatile (e.g., methylarsonic acid and dimethylarsinic acid) compounds [66]. To avoid degradation of arsenic speciation, biological samples should be kept in low temperature. To reduce analyte loss, drying in oven used for the stabilization of samples particularly freeze-drying [67].

To avoid arsenic speciation loss during sampling, the soil–plant-water should be collected in polyethylene flip-top bottle/plastic with lock and/or seal lead. Immediately after collection all of the samples should be kept in freezer until sample preparation for analysis. The soil samples were air-dried, gently crushed and sieved through a 2 mm sieve and used for analysis. Meanwhile the plant sample placed in a oven drier at 40°C until constant weight and then grind and sieve the sample and stored in brown glass bottles in a desiccator in order to avoid exposure to light and moisture until required for analysis. Sample preparation for solid samples generally may include procedures such as mincing, freeze drying, milling, grinding, homogenization, and sieving, followed by extraction. For achieving the best extraction efficiency and reproducibility of arsenic speciation in soil and plant sample, the tested sample must dried and homogenized before extraction. Because, particle size plays a crucial role in the extraction efficiency of As [68]. After sampling the fresh plant sample, it should be kept in freezing (−80°C) to avoid species interchange. Moreover, dry and grind plant and soil sample can store at −20°C up to one year [69].

The most reliable method for preserving natural water samples is, therefore, acidification to pH 2, refrigeration and deoxygenation [70]. Preservation of natural water in polypropylene bottles in refrigerator arsenic species in water is stabile under neutral conditions for a period of 4 months [71]. To increases the stability of dissolved As redox species (As (III) and As(V)) of water sample, the samples to be filtered and stored in darkened polythene containers. For longer preservation, water samples are acidified with HCl [72], HNO_3 [73], H_2SO_4 [74] and H_3PO_4 [75], ascorbic acid [76] ethylene diamine tetraacetic acid [65]. Filtering the sample removes most of the colloidal material and microorganisms; acidification prevents oxidation and precipitation of Fe and Mn hydroxides and EDTA sequesters Fe and inhibits precipitation. Using 10 mM H_3PO_4 as a preserving agent combined with keeping samples at 6°C in dark, As species remain stable for 3 months, even with high concentrations of Fe and Mn.

3.2 Extraction procedures for arsenic speciation

The great challenge of As speciation, as it has been highlighted, is to maintain the original characteristics of species during extraction step. Extraction is the first step for speciation of target As species from their matrix (water, soil, sediment, plant, biological tissue or fluid). Determining an appropriate sample preparation method that provides high extraction efficiency for the arsenic species of interest and prevents inter-conversion between arsenic species can be challenging. To achieve maximum extraction competence of arsenic speciation from solid or liquid matrix, the extraction protocols must have three criteria's, such as (i) the extracting solution must solubilize only the specific form, (ii) reduction of native As (V) to As (III) may not occur during the extraction, and (iii) oxidation of native As (III) to As (V) should not occur [77]. Extraction procedures employ a range of approaches including solid–liquid extraction [78] liquid– liquid extraction, solid phase extraction (SPE) [79] and solid phase microextraction (SPME) [80]. Enhanced techniques such as soxhlet, [69] sonication, [81] pressurized liquid extraction (PLE), [82] microwave-assisted extraction (MWA) [83] and supercritical fluid extraction (SFE) [84] have also been utilized for the determination of As speciation in soil–plant-water sample. Soil-pant-water sample preparation and extraction methods applied for the arsenic speciation analysis are presented in **Table 2**.

3.2.1 Solvent extraction

The solvent extraction technique is often used to determine organic arsenic compounds, especially arsenic compounds in biological samples. The extraction of arsenic substances is usually achieved through mild extraction solvents (ie water, methanol, methanol–water solvent system) and/or rarely uses acetonitrile-water and sequential extraction [15, 103]. Methanol/water mixture 1/1 (v/v) is widely used for the quantification of water-soluble As compounds in environmental samples, followed by centrifugation and filtration [104], while methanol-chloroform or hexane is used in non-polar species [15]. Moreover, extraction with water–methanol (1:1vv^{-1}) had offered easy oxidation of As (III) in basic medium such as soil and the best efficiency was achieved after 20 min of extraction [105]. Extraction efficiency of arsenic species in soil–plant-water samples varied according with the changing the ratio of methanol– water solvent. Nevertheless, the methanol:water extraction solvent ratio of 1:1 provides the best processing and extraction efficiency for the extraction of arsenic species from plant samples, while 1 M phosphoric acid is suitable for soil samples [15]. At the same time, Rahman et al., [98] noticed that addition of extracting agent $NH_4H_2PO_4$ in edible part of spinach had shown similar extraction efficiency of As (III) and As (V) by water, 50% vv^{-1}water/methanol solution on shaking and microwave techniques. However, As(III) was extracted twice as much by the protein extract, indicating that it is a good extractor. Zheng and Hintelmann [106] pointed out that methanol/water mixture is an effective extractant for organic species, and its efficiency for inorganic species drops sharply. The solvent extraction reagent, tetramethylammonium hydroxide (TMAH) in alkaline medium, is also useful for the determination of AB, DMA and inorganic arsenic form a wide variety of biological matrices. In addition, sequential extraction procedure using different solvents (i.e $(NH_4)_2SO_4$, $(NH)_4H_2PO_4$, NH_4-oxalate buffer, KOH and hot water) can effectively extract organic and inorganic arsenic species, namely arsenous acid, arsenic acid, monomethylarsonic acid, dimethylarsinic acid, arsenobetaine, trimethylarsine oxide and glycerol-ribose in both soil and plant [107, 108]. Larios et al. [109] found that orthophosphoric acid followed by graphite block heating at 90°C for 60 min was provided the best conditions for As speciation in plants grown in contaminated environment. The applied extraction solvent led to an extraction efficiency of 80% for samples without species interconversion and recovery of 95% for leaves As speciation of Arsenic (V), As (III), DMA and MMA.

Matrix	Arsenic species	Sample preparation/extraction	Extraction solution	Detection	References
Soil	AS total, AS^{III}, AS^{V}	Shaking/mixing	10 M HCl	HGAAS; XRF	[85]
Soil	AS total, AS^{III}, AS^{V}	Shaking/mixing	10 mM phosphate	HPLC	[86]
Soil	AS total, AS^{III}, AS^{V}	Micro wave heating + Shaking	1 M H_3PO_4 + 0.5 M ascorbic acid (C6H8O6)	LC=UV- HG- AFS	[87]
Soil	AS^{III}, AS^{V}, MMA^{V}, DMA^{V}	Ultrasonic, shaking, Microwave heat	500 mM Phosphate solution	HPLC-HG-AFS	[88]
Soil	AS^{III}, AS^{V}	Filtration. Shaking and water bath heat	[BMIM][PF6]	IL-LLME-ETAAS	[89]
Soil	AS total	Microwave heat	HNO_3 + $HClO_4$+ H_2SO_4	HGAAS	[90]
Soil	AS total, AS^{III}, AS^{V}	Microwave heat	1 M H_3PO_4 + 0.1–1 M ascorbic acid	LC–HG–AFS	[91]
Soil	AS^{III}, AS^{V}	Shaking/mixing	2.5 mM CaCl2	LC–HG–AFS	[21]
Soil	AS^{V}	Shaking/mixing	1 M HCl	XRF	[92]
Soil and plant (chickpea)	AS^{III}, AS^{V}, MMA^{V}, DMA^{V}	Shaking/mixing	Buffer solution (1.5 mM NaH_2PO_4+ 0.2 mM Na_2EDTA + 3 mM $NaNO_3$,+ 10 mM CH_3COONa + 1% C_2H_5OH; pH 6.0)	ICP-MS	[93]
Plant	AS^{III}, AS^{V}, MMA^{V}, DMA^{V}	Shaking/mixing	0.3 M H_3PO_4	HPLC-ICP-MS	[94]
Plant	AS^{III}, AS^{V}, MMA^{V}, DMA^{V}	Shaking/mixing	1% HCOOH	HPLC-ICP-MS-ESI-MS	[95]
Plant	AS^{III}, AS^{V}, MMA^{V}, DMA^{V}	Sonication	2 mM NaH_2PO_4 + 0.2 mM Na_2EDTA (pH 6.0)	HPLC-ICP-MS-ESI-MS.	[95]
Soil and plant	AS^{III}, AS^{V}, MMA^{V}, DMA^{V}, AsC	Shaking/mixing + sonication	CH3OH/H2O 1 + 1 v/v	HPLC-ICP-MS	[96]
Soil and plant	AS^{III}, AS^{V}, MMA^{V}, DMA^{V},	Shaking/mixing + sonication	(a) CH3OH/H2O 1 + 1 v/v; (b) 0.1 M HCl	HPLC, AAS and XANES	[97]
Plant	AS^{III}, AS^{V},	MW-heating	0.33 M sucrose, 50 mM MES, 5 mM EDTA, 5 mM Lascorbate	HPLC-ICP-MS	[98]
Plant	Total As, AS^{III}, AS^{V},	MW-heating	Methanol–water (1: 1) /HNO_3	HPLC-ICP-MS	[99]

Matrix	Arsenic species	Sample preparation/extraction	Extraction solution	Detection	References
Plant	AS^{III}, AS^{V}, MMA^{V}, DMA^{V},	MW-heating	1% (v/v) HNO_3	HPLC-ICP-MS	[100]
Plant	AS^{III}, AS^{V}, MMA^{V}, DMA^{V},	MW-heating	0.01 mol/L TMAH	ETAAS	[101]
Surface/drinking water	AS^{III}, AS^{V}, MMA^{V}, DMA^{V}	Filtration	EDTA	HPLC-ICP-MS	[102]
Sea water	AsIII, AsV, MMA, DMA, AsB, TMAO	Shaking/mixing + ulta-sonication	1% (v/v) HNO_3	HPLC-ICP-MS	[26]

Table 2.
Several soil-water-plant sample preparation/extraction methods for determination of arsenic speciation.

3.2.2 Enzymetic hydrolysis

Biomolecular hydrolysis of complex matrix, enzymes are able to break down specific bonds of the substrate at neutral pH and room temperature, and they allow selective release of the analyte from the sample matrix without chemicalchanges. Enzymes can digest various matrix components, enzyme-assisted reactions usually require several hours of incubation. Microwave-assisted extraction (MAE) is used in combination with the enzyme extraction of pronase E and lipase to effectively extract AsB, As(III), DMA, MMA, and As(V) from seafood, rice, and plants [110, 111]. Viscozyme, was considered the most effective multi-enzyme mixture (consisted of a wide range of carbohydrases, including arabanase, cellulase, glucanase, hemicellulase, and xylanase) useful to extracted arsenic species from algae and terrestrial plant materials [112].

3.2.3 Microwave-assisted extraction

Microwave extraction is a common technique for extracting biological and environmental matrices, which is much faster than traditional Soxhlet extraction procedures. The extraction procedure using dilute acids or organic solvents at low temperatures can be easily achieved in a focused microwave oven. Generally microwave extraction is used to determine inorganic arsenic in food and provided good arsenic speciation extraction efficiencies (generally >90%) for samples of rice and wheat [113]. The method is based on extracting samples with trifluoroacetic acid/H_2O_2 and measuring arsenate by anion exchange HPLC-ICP-MS using aqueous malonic acid as the mobile phase. By using 2 M trifluoroacetic acid assisted with microwave heating for 6 h at 100°C to hydrolyze rice samples, the conversion between As^{III} and As^{V} was also observed and recovered 83, 88, 100, and 93% of fortified arsenite (100 ng As g^{-1}), arsenate (100 ng As g^{-1}), methylarsonic acid (MMA, 50 ng As g^{-1}), and dimethylarsinic acid (DMA, 200 ng As g^{-1}), respectively [114].

3.2.4 Solid phase extraction

Solid phase extraction (SPE) method is an efficient extraction technique for arsenic speciation from complex environmental and biological matrices. The principle mechanism of SPE is partitioning sorbent and sample matrix phase and may include simple adsorption, chelation, ion exchange or ion-pair solid phase extraction. In recent years, the techniques gaining popularity for As speciation because of its simple operation offers acceptable recovery and pre-concentration efficiency, lower reagent consumption and offer effective combination ability with different on-line and off-line As detection systems.

3.2.4.1 Conventional sorbent

Several conventional sorbent (i.e ion exchange resin, glass and modified mesoporous silica) based protocols have been developed for inorganic As speciation. To avoid inter-conversion of arsenic species, extraction with anion exchange cartridges prior to the inductively coupled plasma sector field mass spectrometric (ICP-SF-MS) becomes an efficient technique. During on- site separation and speciation of inorganic arsenic (As (III) and As (V)) from high arsenic- groundwater and ferrihydrite removal anaerobic arsenics species, anion-exchange resin (AG 1-X8) adsorbed As(V) in acetate form, while no adsorption to As(V)/As(III) in chloride form [115]. A dual-sorbent SPE protocol, in which the sorbent is composed of strong basic anion exchange (SBAE) resin and hydrate iron oxide particles integrated HY resin, has been adopted successfully for the retention of inorganic arsenic species As (V) and As(III) simultaneously [116]. On-line continuous leaching extraction method is also effective for speciation of bio-accessible As species in plant [108].

3.2.4.2 Functional nanomaterials extractant

The modern technological invention of nanomaterials such as Nanofibers [117], magnetic nanoparticles [118], metal hydroxide precipitation [119], and nano-TiO_2 colloid [120] has offered selective and efficient extraction techniques for As speciation from different matrix. Like ammonium pyrroine- dithiocbonate (APDC) have excellent selectivity of As (III) from ground water samples [117]. Moreover, yttrium hydroxide precipitate layer coated cellulose fiber is used as extracting material, [119] of As (III) and As (V) at acidic condition. Multi-wall carbon nanotubes (MWCNTs) modified with branched cationic polyethyleneimine(BPEI)is also proved to be excellent adsorbent with favorable selectivity toward adsorption of As(V) [121] in combined with sequential injection technique. Nano particle TiO_2 colloid has effectively extracted ultra-trace As from environmental water sample without agglomeration [120]. Besides, As (III) and As(V)from aqueous solution can be effectively extracted by hematite-coated Fe_3O_4 particles. Moreover, due to the fact simple and rapid separation capacity of As species, magnetic extraction techniques also gaining much popularity day by day.

3.2.4.3 Multi-sorbent based SPE procedure

A combined SPE procedure for arsenic speciation developed by using three molecular recognition technology (MRT) gel resins, which includes strong base anion exchange (SBAE) and two hybrid (HY) resins, HY–Fe and HY–AgCl, This methods has constructed for simultaneously extraction of four water-soluble arsenic species: arsenite, arsenate, monomethylarsonic acid (MMA) and dimethylarsinic acid (DMA) and retain in the SPE columns and separatedly eluted by using different elution [115].

3.2.5 Liquid–liquid extraction (LLE)

A liquid–liquid extraction generally used dodecane modified with 4% dodecanol containing Aliqua t336 as the extractant has been developed for the extraction of arsenic species in environmental matrices [122]. Here only As (V) is quantitatively transported to organic phase but no transport of As(III) takes place. A rapid in-situ liquid–liquid micro-extraction procedure has been developed for successfully determination of As (III) and As(V) in water samples, food salts, and total As in biological samples [123].

4. Derivitization of total arsenic

The vital challenge in element speciation is to determine each form independently without interference from other species. Because arsenic could complex with certain derivatizing agents, that hampered the detection. The derivitization process consists of two steps for example. i) reduction of AsV to AsIII and ii) convert to arsine (AsH_3) [124]. During measurement, the inert g N_2 is pushed by hydride generation (HG) step, reaches the atomic absorption spectrophotometer or ICP-MS and finally produced arsines [125]. The main limitation of derivitization is that it only limits the formation of volatile arsines materials, but it is difficult to separate un-derivatized arsenic species (eg. AsB, AsC, or arsenosugars), using conventional reversed phase liquid chromatography and almost impossible by spectrophotometry or mass spectrometry [126]. In addition, the derivitization process requires control condition In addition, the derivitization process requires control condition. This technique largely depends on the type and concentration of the sample matrix. To overcome obstacle, sodium borohydride is now commonly used for the hydrides synthesis [18]. Arsenic speciation after derivatization can be overcome by combining couple technique with detection techniques such as LC–MS/MS retention in liquid chromatography and ionization in mass spectrometry. Under such circumstances, the hyphenated technique is the most reliable because it includes several facilities like high sensitivity, good reproducibility, short analysis time and reduced risk for species transformation [18].

5. Separation techniques of arsenic speciation

Usually, two main techniques, namely chromatography (gas and liquid) and capillary electrophoresis are used to separate arsenic from various complex matrices [65]. Based on the complexity of As compounds, sometimes two technologies are introduced simultaneously or cumulatively.

5.1 Liquid chromatography

Liquid chromatography generally provides excellent possibilities for the separation of environmental and biological samples [18]. Various commonly used liquid chromatography techniques are high performance liquid chromatography (HPLC), ion chromatography and ion interaction chromatography [127]. Chromatographic separation can be performed by using ion exchange columns to separate metal ions directly or by adsorption (reverse phase or normal phase) liquid chromatography (if the metal species are complexed with organic ligands). Liquid chromatography is connected to many other detection systems, such as ICP-MS, HG-AFS, HG-AAS and GF-AAS [65]. Several liquid chromatography techniques can be used for the organic and inorganic As species, as follows:

5.1.1 Anion exchange liquid chromatography

Anion exchange chromatography is commonly used for speciation analysis of arsenic in environmental and biological samples. The anionic nature of arsenic species is different (at neutral pH, arsenic acid As(V), monomethylarsonic acid (MMA) and dimethylarsonic acid (DMA) are deprotonated, but As(III)) exists) to make anion exchange chromatography suitable for their separation. Anion exchange chromatography has been used to analyze many arsenic compounds, including As(III), As(V), MMA, DMA, arsenobetaine (AsB), arsenocholine (AsC), oxoarsenic sugar (oxoAsS), thiosulfate Arsenic sugar (thioAsS) and benzene arsenic [27, 28]. The most commonly used column for arsenic speciation analysis is a strong anion exchange column, such as PRP-X100. Generally, the As form of the matrix separated by anion exchange chromatography techniques is detected by inductively coupled plasma mass spectrometry (ICP-MS) and electrospray ionization tandem mass spectrometry (ESI-MS/MS).

5.1.2 Cation exchange liquid chromatography

Cation exchange chromatography works similarly to anion exchange, except that the stationary phase is negatively charged to interact with the cation analyte. However, in cation exchange liquid chromatography, the separation speed of As species is relatively fast. The retention of arsenicals is directly related to the strength of its cationic charge: positively charged analytes have stronger retention. Cation exchange chromatography is commonly used for speciation analysis of positively charged As compounds, such as AsB, AsC, trimethylarsenic oxide (TMAO) and tetramethylarsenic ion (TMA) [15].

5.1.3 Reverse phase liquid chromatography

Reversed-phase chromatography is the most common HPLC separation technique used to separate compounds that are less hydrophobic or polar. In particular, reversed-phase liquid chromatography is particularly suitable for the analysis of arsenic lipids, including arsenic-containing hydrocarbons, fatty acids, phospholipids, phosphatidylcholines, fatty alcohols, and phosphatidylethanolamines of biological samples [24].

5.1.4 Ion pair chromatography

Ion pair chromatography can separate ions and neutral analytes using popular reversed-phase chromatography. It has been widely used for arsenic speciation analysis of various substrates. The reagent of ion pair chromatography reagent comprises with two groups a charged group for interaction with the analyte and a hydrophobic region for interrelates with the stationary phase. Usually, tetraalkylammonium, tetrabutylammonium and tetraethylammonium are used as the ion pair reagents for the separation of anionic and neutral arsenic species, and alkyl sulfonates, such as hexanesulfonic acid and 1-pentane sulfonic acid, for cationic and neutral arsenic species. Two most commonly used organic modifiers, methanol and acetonitrile are added to the mobile phase to decrease retention time [15].

5.1.5 Hydrophilic interaction liquid chromatography

Hydrophilic Interaction Chromatography (HILIC) is an important substitute to RP-HPLC separations of polar compounds. Although the stationary phase is polar, HILIC can separate neutral, cationic and anionic species simultaneously. HILIC has great potential to separate more arsenic species in a single analysis. This separation technique is more useful for organoarsenicals. Xie et al. [128] successfully detected nine kinds of organoarsenicals (I,e MMA, DMA, AsB, AsC, TMAO, phenylarsonic acid (PAA), phenylarsine oxide (PAO), 4-hydroxy-3-nitrophenylarsonic acid (Roxarsone), and 4-aminophenylarsonic acid (p-arsanilic acid, ASA) using a zwitterionic HILIC column followed by ICP-MS/ ESI-MS detection.

5.1.6 Size exclusion chromatography

Size-exclusion chromatography (SEC), also known as molecular sieve chromatography, is a chromatographic method in which molecules in solution are separated by their size, and in some cases molecular weight. SEC is very effective for analysis of arsenic interactions with large molecules or macromolecular complexes such as arsenic-protein binding, humic acid-arsenic complexesand industrial polymers. SEC commonly used to separate protein-bound arsenic from free arsenic [129]. This separation method is expensive and useful for biological samples.

5.1.7 Multidimensional chromatography

Combining a variety of chromatographic columns and separation modes, try to separate a series of arsenic substances. Multidimensional separation has been performed offline or online. These usually involve a cation exchange column and an anion exchange column connected by a switching valve. This combination allows separation of cationic and anionic arsenic species. Applications were demonstrated for the determination of water-soluble arsenic species [20].

5.2 Capillary electrophoresis

Capillary electrophoresis separates As species according to the electrophoretic mobility related to the charge, viscosity, and atomic radius of the molecule, which is controlled by the composition, concentration, and pH of the buffer. This method is applicable for all type of soil–plant-water samples. Capillary electrophoresis can be used in many different detection systems but the most common is ICP-MS [15]. Although, Although capillary electrophoresis separation is simple, cost-effective, fast analysis and a certain degree of matrix independence, the additional complexity of coupling with the detection system makes CE a less common As speciation analysis method.

6. Detection systems of arsenic species

Several sensitive and element techniques can be used for the detection of arsenic species. Various detection techniques are: atomic mass spectrometry, molecular mass spectrometry, optical spectrometry, X-ray methods and others (voltammetry, potentiometry, conductometry and spectrophotometry), which provide different level of specificity, cost effectiveness and detection limits [21]. Detection methods applied for the arsenic speciation analysis of soil–plant-water samples are assembled in **Table 3**.

ICP-MS is the most commonly used technique for the detection of multiple arsenic species because of its high sensitivity, high selectivity and wide dynamic range. The coupling of chromatography to ICP-MS has several benefits due to the compatibility of the mobile phase with the behavior of the plasma torch and the carefully determined quality inspection interference. Various techniques have been developed to eliminate or reduce isobaric interference in the detection of arsenic by mass-to-charge ratio. Recently, compared with the traditional single quadrupole ICP-MS, the combination of ICP and triple quadrupole tandem mass spectrometry (ICP-QQQ) helps to eliminate isobaric interference, reduce background, and improve selectivity [156]. Quantification is performed by preparing standard solutions of commercially available substances, such as iAs(III), iAs(V), MA, DMA, and AB. It is generally believed that arsenate is used to calibrate anionic substances, and arsenobetaine is used to calibrate cationic substances [157]. DMA is considered to be the most suitable calibration standard for arsenic lipid quantification [158]. The sensitivity of ICPMS to detect arsenic is limited by its relatively high ionization potential. In order to compensate for this effect, various methods have been used,

Matrix	Arsenic species	Detection techniques	Coupled with	References
Atomic mass spectrometry				
Soil	AS^{III}, AS^{V}	ICP = MS	—	[130]
Water	AS^{III}, AS^{V}, MMA^{V}, DMA^{V}			[131]
Soil-water-plant	Total As, AS^{III}, AS^{V},			[132]
Water	AS^{III}, AS^{V}	ICP-SFMS	—	[133]
Water	AS^{III}, AS^{V}, MMA^{V}, DMA^{V}, AsB AsC	ICP = MS	HPLC	[134]
Plant	AS^{III}, AS^{V}, MMA^{V}, DMA^{V}, AsB			[135]
Soil	AS^{III}, AS^{V}			[136]
Soil	AS^{III}, AS^{V}, MMA^{V}, DMA^{V}	ICP-SFMS	HPLC	[137]
Plant	AS^{III}, AS^{V}, MMA^{V}, DMA^{V}			[138]
Water	AS^{III}, AS^{V}, MMA^{V}, DMA^{V}			[139]
Soil	MMA^{V}, DMA^{V}	ICP = MS	GC	[140]
Soil	AS^{III}, AS^{V}, MMA^{V}, $DMA^{V,}$	ICP = MS	CE	[137]
Molecular mass spectrometry				
Soil	PA and AA	ESI-qMS	HPLC	[141]
Soil -water	PA and AA	ESI-TOF-MS	=	[142]
Soil–plant	AS^{III}, AS^{V}, MMA^{V}, DMA^{V}, MMMTA, DMMTA, DMDTA	ESI-qTOF-MS	=	[143]
Soil–plant	AS^{III}, AS^{V}, MMA^{V}, DMA^{V}, MMMTA, DMMTA, DMDTA	ESI-qTOF-MS	HPLC	[143]
Soil	AS^{III}, AS^{V}, N-AHPAA, 3-AHPAA	ESI-triple quad-MS	=	[144]
Plant	Arsenolipids	ESI-triple quad-MS	HPLC	[145]
Water	AS^{III}, AS^{V}, MMA^{V}, DMA^{V}, TMAO	ESI-Orbitrap-MS	=	[146]
Plant	Arsenic peptides	ESI-IT-MS	HPLC	[147]
Plant	PA and AA	EI-MS	GC	[141]
Optical spectroscopy				
Plant	AS total, AS^{III}, AS^{V}	GF-AAS	=	[148]
Soil	AS total, AS^{III}, AS^{V}	HG-AAS	=	[85]
Water	AS^{III}, AS^{V}, MMA^{V}, DMA^{V}	HG-AAS	HPLC	[127]
Soil–plant	AS^{III}, AS^{V}, MMA^{V}, DMA^{V}			[69]
Soil	AS total, AS^{III}, AS^{V}	HG-AFS	=	[87]
Soil	AS^{III}, AS^{V}, MMA^{V}, DMA^{V}	HG-AFS	HPLC	[88]
Plant	AS^{III}, AS^{V}, MMA^{V}, DMA^{V}			[149]

Matrix	Arsenic species	Detection techniques	Coupled with	References
X-ray methods				
Soil and Plant	AS^{III}, AS^{V}, MMA^{V}, DMA^{V}	XANES		[150]
Soil	AS^{III}, AS^{V}	EXAFS		[151]
Soil	AS^{III}, AS^{V}	STXM		[152]
Soil	AS^{V}	XPS		[153]
Others				
Water	AS^{III}, AS^{V},	Voltammetry	=	[154]
Water	AS^{V}	Potentiometry	=	[155]
Water	AS^{III}, AS^{V}	Spectrophotometry	HPLC	[126]

Aqueous phenylarsonic acid (PA); o-arsanilic acid (AA); N-acetyl-4-hydroxy-m-arsanilic acid (N-AHPAA), 3-amino-4-hydroxyphenylarsonic acid (3-AHPAA).

Table 3.
Examples of detection systems for arsenic speciation analysis of soil-water-plant samples.

including adding supplemental methanol or ethanol solution to the spray chamber [159] or after the column via the T-piece [91], and the use of correction response factors. Finally, internal standardization was used to overcome the non-spectral matrix effects and instrumental drift [160].

Recently, molecular mass spectrometry is considered as a forward-looking technique for arsenic speciation analysis, especially for the detection of new organic arsenic species, such as thioarsenosugar [21, 161] and arsenolipids [21, 60, 162]. In this detection technique, the purified part of the extractable sample is introduced by electrospray ionization (ESI), and then mass spectrometry is combined with liquid chromatography. Generally, for the As forms, a simple single quadrupole mass analyzer is used, while tandem mass spectrometry is used for precise structure determination, whether it is a "spatial" triple quadrupole or a quadrupole time combination, or a "time" and Orbitrap system [21]. However, it has been recognized that ESI-MS analysis lacks selectivity for complex matrices, and quantification is more difficult than ICP-MS [163]. Therefore, the most powerful setting for arsenic speciation analysis that combines atomic and mass spectrometers is used as the detector of the same chromatographic system [21, 60, 145].

The optical spectroscopy technique such as atomic absorption spectroscopy (AAS) and atomic fluorescence spectroscopy (AFS) is popular to researcher as an attractive alternative to mass spectrometry. Due to the low purchase and operation cost, high speed, low consumption of organic solvents, high enrichment coefficient, combined with hydride generation provides high sensitivity and reduced matrix effect, this technology has been applied to the determination of arsenic species in environmental samples. Moreover, hydride generation systems (HG-AAS and HG-AFS) facilitate a direct measurement of the more As. Graphite furnace atomic absorption spectroscopy can be an independent facility and does not require AsH3 because of the low level of interference [18]. In fact, the optical spectroscopy is an effective technique, when combined with different separation techniques and chemical modifiers, iAs(III), iAs(V), MA, DMA and TMAO, can be identified, and significant hydride generation of arsenosugars [164] and thioarsenates can be observed [21]. Nevertheless, HGAAS and HG-AFS are mainly used for water samples [150], sediment extracts and soil, [165] and plants [97] mainly contain inorganic arsenic. These techniques are also applicable to biological substrates and more stubborn arsenic Analysis [166].

X-ray method is an important technique for morphological analysis of environmental samples, which can record raw data about the chemical environment of arsenic atoms in situ without sample preparation. X-ray atomic absorption spectroscopy (XAS) is generally divided into two regions: X-ray absorption near edge structure (XANES) and extended X-ray absorption fine structure (EXAFS). These technologies are mainly used to directly detect solid samples, including sediments, [167] soil [97, 168] and plants [96, 97, 169]. Both XANES and EXAFS have studied abiotic matrices to measure arsenic redox status and geochemical correlation.

7. Accuracy evaluation of speciation methods

In order to obtain precise analytical information about the bioavailability and toxicity of arsenic in the environmental process, it is necessary to carefully consider any possible sources of error during analysis and validate the data. To avoid or minimize the impact of species changes and ensure the reliability and quality of speciation data, mass balance ratio, extraction efficiency, column recovery of arsenic species during separation and standard reference materials quality need to be tracked. The main difficulty of specific analysis of arsenic may occur in the sample preparation stage and species stability. Mass balance data provides information about the distribution of elements in each analysis step (extraction, separation, and species detection) and quantitatively determine the fate of arsenic during speciation [170]. The extraction efficiency can provide some important information about the extraction procedure, the polarity of the extracted species, and help to select the best extraction solvent and separation system. It helps to establish a non-toxic, effective and simple extraction procedure for arsenic speciation analysis. Column recovery is an important aspect of any separation technique. It is critical to eliminate loss and to ensure there is no cross-contamination between analyses. The column recovery compares the total arsenic concentration with the sum of the detected substances, which can provide information about the elution and retention of the analyte. In fact, depending on the type of sample and the concentration of the arsenic species, the column recovery rate of the arsenic species has great variability [171]. The column recovery also affected by the extraction solvent of the column. It is difficult to evaluate the mutual transformation of arsenic species in the actual sample in the column, which is related to the individual arsenic standard. However, the lack of available standards for new arsenic species is the main challenge in studying the inter-conversion of arsenic compounds during separation [15]. For accurate method validation of arsenic speciation, the use of standard reference materials (SRM) and certified reference materials (CRM) is essential. With reference or certified values available, SRM and CRM can be used to test and verify the accuracy of the method. In order to verify the arsenic speciation analysis methods of environmental samples, different types of soil and sediments, natural waters, marine and terrestrial plants and other biological samples are used as reference samples. It should be noted that a single SRM or CRM could not be used to verify method calibration and results [15]. SRM 1640 (NIST) is commonly used to check calibration curves for trace elements in water. The type of CRM used depends on the sample matrix and the type of arsenic studied [1].

8. Conclusion

Arsenic pollution is a universal problem. The form of arsenic in soil, water, and plants play an important role in understanding arsenic exposure, metabolism and environmental arsenic cycle, and food chain. A crucial requirement for obtaining reliable speciation information is to maintain the concentration of the original chemical species in the sample prior to analysis. In order to determine the total element concentration, the main considerations for sample collection and storage are to prevent contamination and minimize the loss of trace analytes. Research on simple and efficient extraction procedures that use less or non-toxic solvents is very urgent for better arsenic speciation In the case of speciation analysis, the concentration of individual species of the element must be constant through sample handling and processing. Therefore, the time between the extraction procedure and the analysis must be as short as possible to avoid interconversion between species. The selection of extraction and sample preparation methods must be complementary and compatible with the separation method in order to perform qualitative and quantitative analysis of arsenic species and its concentration. It may require a combination of multiple extraction methods and multiple separation techniques to achieve a comprehensive arsenic speciation analysis. Several techniques have been used to study arsenic speciation, each with its advantages and disadvantages. However, research efforts are still needed to develop cheap, fast, sensitive, and reproducible methods for arsenic species that can work at low detection limits. However, research efforts are still needed to develop cheap, fast, sensitive, and reproducible methods for arsenic species that can work at low detection limits. In addition, in order to find a unified analysis protocol i.e. at least for the more common matrices, for the prevalent and unidentified arsenic species, advanced investigations and routine measurements are necessary.

Conflict of interests

The authors declare that they have no conflicting interests.

Author details

Mohammed Zia Uddin Kamal* and Md. Yunus Miah
Department of Soil Science, Bangabandhu Sheikh Mujibur Rahman Agricultural University (BSMRAU), Gazipur, Bangladesh

*Address all correspondence to: zia_ssc@yahoo.com; zia@bsmrau.edu.bd

References

[1] Sadee B, Foulkes ME, Hill SJ. Coupled techniques for arsenic speciation in food and drinking water: a review. Journal of Analytical Atomic Spectrometry. 2015; 30(1): 102-18.

[2] Khalid S, Shahid M, Niazi NK, Rafiq M, Bakhat HF, Imran M, Abbas T, Bibi I, Dumat C. Arsenic behaviour in soil-plant system: Biogeochemical reactions and chemical speciation influences. InEnhancing cleanup of environmental pollutants 2017 (pp. 97-140). Springer, Cham.

[3] Chandrakar V, Naithani SC, Keshavkant S. Arsenic-induced metabolic disturbances and their mitigation mechanisms in crop plants: A review. Biologia. 2016 May 1; 71(4): 367-77.

[4] Abbas G, Murtaza B, Bibi I, Shahid M, Niazi NK, Khan MI, Amjad M, Hussain M. Arsenic uptake, toxicity, detoxification, and speciation in plants: physiological, biochemical, and molecular aspects. International journal of environmental research and public health. 2018; 15(1): 59.

[5] Huq SI, Alam MD. A handbook on analyses of soil, plant and water. BACER-DU, University of Dhaka, Bangladesh. 2005; 246.

[6] Chakraborty S, Alam MO, Bhattacharya T, Singh YN. Arsenic accumulation in food crops: a potential threat in Bengal delta plain. Water Quality, Exposure and Health. 2014 Dec 1; 6(4): 233-46.

[7] Rahman MM, Asaduzzaman M, Naidu R. Arsenic exposure from rice and water sources in the Noakhali district of Bangladesh. Water Quality, Exposure and Health. 2011 Jun 1; 3(1): 1-0.

[8] Rahman MA, Kadohashi K, Maki T, Hasegawa H. Transport of DMAA and MMAA into rice (*Oryza sativa* L.) roots. Environmental and Experimental Botany. 2011 Aug 1; 72(1): 41-6.

[9] Rauf MA, Hakim MA, Hanafi MM, Islam MM, Rahman GK, Panaullah GM. Bioaccumulation of arsenic (As) and phosphorous by transplanting Aman rice in arsenic-contaminated clay soils. Australian Journal of Crop Science. 2011 Jan; 5(12): 1678-84.

[10] Punshon T, Jackson BP, Meharg AA, Warczack T, Scheckel K, Guerinot ML. Understanding arsenic dynamics in agronomic systems to predict and prevent uptake by crop plants. Science of the Total Environment. 2017 Mar 1; 581:209-20.

[11] FAO/WHO (Session TS). Joint FAO/ WHO Food Standards Programme Codex Committee on Contaminants in Foods, 8th Session. Codex. 2014 Dec; 150(2014/8). World Health Organization, Geneva

[12] US Food and Drug Administration. Questions & Answers: Arsenic in Rice and Rice Products; US. FDA: Silver Spring, MD, USA, 2015

[13] Bakhat HF, Zia Z, Fahad S, Abbas S, Hammad HM, Shahzad AN, Abbas F, Alharby H, Shahid M. Arsenic uptake, accumulation and toxicity in rice plants: possible remedies for its detoxification: a review. Environmental Science and Pollution Research. 2017 Apr 1; 24(10): 9142-58.

[14] Sun Y, Liu G, Cai Y. Thiolated arsenicals in arsenic metabolism: occurrence, formation, and biological implications. Journal of Environmental Sciences. 2016 Nov 1; 49:59-73.

[15] Reid MS, Hoy KS, Schofield JR, Uppal JS, Lin Y, Lu X, Peng H, Le XC. Arsenic speciation analysis: A review with an emphasis on chromatographic

separations. TrAC Trends in Analytical Chemistry. 2020 Feb 1; 123:115770.

[16] Moe B, Peng H, Lu X, Chen B, Chen LW, Gabos S, Li XF, Le XC. Comparative cytotoxicity of fourteen trivalent and pentavalent arsenic species determined using real-time cell sensing. Journal of Environmental Sciences. 2016 Nov 1; 49:113-24.

[17] Ebert F, Leffers L, Weber T, Berndt S, Mangerich A, Beneke S, Bürkle A, Schwerdtle T. Toxicological properties of the thiolated inorganic arsenic and arsenosugar metabolite thio-dimethylarsinic acid in human bladder cells. Journal of Trace Elements in Medicine and Biology. 2014 Apr 1; 28(2): 138-46.

[18] Rukh S, Akhtar MS, Memon M, Mehmood A, Imran M. An overview of arsenic extraction and speciation techniques in soil and water. Amer. Chem. Sci. J. 2015; 6:1-5.

[19] Templeton DM, Ariese F, Cornelis R, Danielsson LG, Muntau H, van Leeuwen HP, Lobinski R. Guidelines for terms related to chemical speciation and fractionation of elements. Definitions, structural aspects, and methodological approaches (IUPAC Recommendations 2000). Pure and applied chemistry. 2000 Aug 31; 72(8): 1453-70.

[20] Gong Z, Lu X, Ma M, Watt C, Le XC. Arsenic speciation analysis. Talanta. 2002 Aug 16; 58(1): 77-96.

[21] Ardini F, Dan G, Grotti M. Arsenic speciation analysis of environmental samples. Journal of Analytical Atomic Spectrometry. 2020; 35(2): 215-37.

[22] Ng JC, Wang J, Shraim A. A global health problem caused by arsenic from natural sources. Chemosphere. 2003 Sep 1; 52(9): 1353-9.

[23] Caumette G, Koch I, Reimer KJ. Arsenobetaine formation in plankton: a review of studies at the base of the aquatic food chain. Journal of Environmental Monitoring. 2012; 14(11): 2841-53.

[24] Taylor V, Goodale B, Raab A, Schwerdtle T, Reimer K, Conklin S, Karagas MR, Francesconi KA. Human exposure to organic arsenic species from seafood. Science of the Total Environment. 2017 Feb 15; 580:266-82.

[25] Jackson BP. Fast ion chromatography-ICP-QQQ for arsenic speciation. Journal of analytical atomic spectrometry. 2015; 30(6): 1405-7.

[26] Park MK, Choi M, Kim L, Choi SD. An improved rapid analytical method for the arsenic speciation analysis of marine environmental samples using high-performance liquid chromatography/inductively coupled plasma mass spectrometry. Environmental monitoring and assessment. 2019 Aug; 191(8): 1-4.

[27] Liu Q, Lu X, Peng H, Popowich A, Tao J, Uppal JS, Yan X, Boe D, Le XC. Speciation of arsenic–a review of phenylarsenicals and related arsenic metabolites. TrAC Trends in Analytical Chemistry. 2018 Jul 1; 104:171-82.

[28] Reis VA, Duarte AC. Analytical methodologies for arsenic speciation in macroalgae: a critical review. TrAC Trends in Analytical Chemistry. 2018 May 1; 102:170-84.

[29] Pfeifer H R, Beatrizotti G, Berthoud J, Rossa M D, Girardet A, Jäggli M, Lavanchy J C, Reymond D, Righetti G, Schlegel C, Schmit V, Temgoua E. Natural arsenic-contamination of surface and ground waters in Southern Switzerland (Ticino). *Bull. Appl. Geol.* 2002; 7:81-103.

[30] Jang YC, Somanna Y, Kim H. Source, distribution, toxicity and remediation of arsenic in the environment–a review. Int J Appl Environ Sci. 2016; 11(2): 559-81.

[31] Onishi, H, & Wedepohl KH. Handbook of geochemistry. 1969; *Vol. II-2*. Springer: Berlin, Germany.

[32] Shahid M, Shamshad S, Rafiq M, Khalid S, Bibi I, Niazi NK, Dumat C, Rashid MI. Chromium speciation, bioavailability, uptake, toxicity and detoxification in soil-plant system: a review. Chemosphere. 2017 Jul 1; 178:513-33.

[33] Niazi NK, Singh B, Shah P. Arsenic speciation and phytoavailability in contaminated soils using a sequential extraction procedure and XANES spectroscopy. Environmental science & technology. 2011 Sep 1; 45(17): 7135-42.

[34] Baig JA, Kazi TG, Shah AQ, Kandhro GA, Afridi HI, Khan S, Kolachi NF. Biosorption studies on powder of stem of *Acacia nilotica*: removal of arsenic from surface water. Journal of hazardous materials. 2010 Jun 15; 178(1-3): 941-8.

[35] Sadiq M. Arsenic chemistry in soils: an overview of thermodynamic predictions and field observations. Water, air, and soil pollution. 1997 Jan; 93(1): 117-36.

[36] Koyama T. Arsenic in soil-plant system. *Nippon Dojo Hiryogaku Zasshi*. 1975; 46:491-502.

[37] Bhattacharya P, Jacks G, Frisbie S H, Smith E, Naidu R, Sarkar B. Arsenic in the environment: a global perspective. in Sarkar, B. (Eds.). Heavy Metals in the Environment. Marcel Dekker. Inc. New York: 2002. p 147-215.

[38] Rhine E D, Phelps C D, Young L Y. Anaerobic arsenite oxidation by novel denitrifying isolates. *Environ. Microbiol.* 2006; 8: 899-908.

[39] Mandal B K, Suzuki K T. Arsenic round the world: a review. *Talanta*. 2002; 58: 201-235.

[40] Sadiq M, Zaidi TH, Mian AA. Environmental behavior of arsenic in soils: Theoretical. Water, Air, & Soil Pollution. 1983 Nov; 20(4): 369-77.

[41] Mahimairaja S, Bolan NS, Adriano DC, Robinson B. Arsenic contamination and its risk management in complex environmental settings. Advances in Agronomy. 2005 Jan 1; 86:1-82.

[42] Marquez EB, Gurian PL, Barud-Zubillaga A, Goodell PC. Correlates of arsenic mobilization into the groundwater in El Paso, Texas. Air Soil Water Res 2011; 4:19-29

[43] Rahaman S, Sinha AC, Mukhopadhyay D. Effect of water regimes and organic matters on transport of arsenic in summer rice (*Oryza sativa* L.). Journal of Environmental Sciences. 2011 Apr 1; 23(4): 633-9.

[44] Williams LE, Barnett MO, Kramer TA, Melville JG. Adsorption and transport of arsenic (V) in experimental subsurface systems. Journal of Environmental Quality. 2003 May; 32(3): 841-50.

[45] Mishra S, Mattusch J, Wennrich R. Accumulation and transformation of inorganic and organic arsenic in rice and role of thiol-complexation to restrict their translocation to shoot. Sci. Rep. 2017, 7, 40522.

[46] Ayangbenro AS, Babalola OO. A new strategy for heavy metal polluted environments: a review of microbial biosorbents. International journal of environmental research and public health. 2017 Jan; 14(1):94.

[47] Shahid N, Zia Z, Shahid M, Faiq Bakhat H, Anwar S, Mustafa Shah G, Rizwan Ashraf M. Assessing Drinking Water Quality in Punjab, Pakistan. Polish J. Environ. Stud. 2015, 24, 2597-2606.

[48] Polizzotto ML, Harvey CF, Li G, Badruzzman B, Ali A, Newville M, Sutton S, Fendorf S. Solid-phases and desorption processes of arsenic within Bangladesh sediments. Chemical Geology. 2006 Apr 16; 228(1-3): 97-111.

[49] Wilkin RT, Wallschläger D, Ford RG. Speciation of arsenic in sulfidic waters. Geochemical Transactions. 2003; 4(1): 1-7.

[50] Cabon JY, Cabon N. Speciation of major arsenic species in seawater by flow injection hydride generation atomic absorption spectrometry. Fresenius' journal of analytical chemistry. 2000 Oct; 368(5): 484-9.

[51] Nriagu J O. Arsenic in the environment. Wiley. New York: 1994. p. 448

[52] Adriano DC. Arsenic. In Trace Elements in Terrestrial Environments; Springer: Berlin, Germany, 2001; pp. 219-261.

[53] Catarecha P, Segura MD, Franco-Zorrilla JM, García-Ponce B, Lanza M, Solano R, Paz-Ares J, Leyva A. A mutant of the Arabidopsis phosphate transporter PHT1; 1 displays enhanced arsenic accumulation. Plant Cell, 2007; 19:1123-1133.

[54] Shin H, Shin HS, Dewbre GR, Harrison MJ. Phosphate transport in Arabidopsis: Pht1; 1 and Pht1;4 play a major role in phosphate acquisition from both low- and high-phosphate environments. Plant Journal, 2004; 39:629-642.

[55] Wu Z, Ren H, McGrath SP, Wu P, Zhao FJ. Investigating the contribution of the phosphate transport pathway to arsenic accumulation in rice. Plant Physiology, 2011; 157:498-508.

[56] Ma JF, Yamaji N, Mitani N, Xu XY, Su YH, McGrath SP, Zhao FJ. Transporters of arsenite in rice and their role in arsenic accumulation in rice grain. Proceedings of the National Academy of Sciences. 2008 Jul 22; 105(29): 9931-5.

[57] Kumar S, Dubey RS, Tripathi RD, Chakrabarty D, Trivedi PK. Omics and biotechnology of arsenic stress and detoxification in plants: current updates and prospective. Environment international. 2015 Jan 1; 74:221-30.

[58] Liu WJ, Wood BA, Raab A, McGrath SP, Zhao FJ, Feldmann J. Complexation of arsenite with phytochelatins reduces arsenite efflux and translocation from roots to shoots in Arabidopsis. Plant Physiology. 2010 Apr 1; 152(4): 2211-21.

[59] Ye WL, Wood BA, Stroud JL, Andralojc PJ, Raab A, McGrath SP, Feldmann J, Zhao FJ. Arsenic speciation in phloem and xylem exudates of castor bean. Plant physiology. 2010 Nov 1; 154(3): 1505-13.

[60] Raab A, Williams PN, Meharg A, Feldmann J. Uptake and translocation of inorganic and methylated arsenic species by plants. Environmental Chemistry. 2007 Jul 13; 4(3): 197-203.

[61] Huang JH, Fecher P, Ilgen G, Hu KN, Yang J. Speciation of arsenite and arsenate in rice grain–Verification of nitric acid based extraction method and mass sample survey. Food chemistry. 2012 Jan 15; 130(2): 453-9.

[62] Abedin MJ, Feldmann J, Meharg AA. Uptake kinetics of arsenic species in rice plants. Plant physiology. 2002 Mar 1; 128(3): 1120-8.

[63] Feldmann J. R. Cornelis, J. Caruso, H. Crews and K. Heumann. Handbook of element speciation techniques and methodology. John Wiley & Sons, 2005. Chichester, ISBN: 0-471-49214-0. 634+ xii pp.49

[64] Moreda-Pi˜neiro, E. Alonso-Rodr'ıguez, A. Moreda-Pi˜neiro,C.

Moscoso-P'erez, S. Muniategui-Lorenzo, P. L'opez-Mah'ıa,; D. Prada-Rodr'ıguez and P. Bermejo-Barrera, Anal. Chim. Acta, 2010, 679, 63-73.

[65] Rajaković LV, Todorović ŽN, Rajaković-Ognjanović V, Onjia A. Analytical methods for arsenic speciation analysis. Journal of the Serbian Chemical Society. 2013; 78(10): 1461-79.

[66] Bentley R, Chasteen TG. Microbial methylation of metalloids: arsenic, antimony, and bismuth. Microbiology and molecular biology reviews. 2002 Jun; 66(2): 250.

[67] Mester Z, Sturgeon RE. Sample preparation for trace element analysis. Elsevier; 2003 Dec 16.

[68] Amaral CD, Nóbrega JA, Nogueira AR. Sample preparation for arsenic speciation in terrestrial plants—a review. Talanta. 2013 Oct 15; 115:291-9.

[69] Mir KA, Rutter A, Koch I, Smith P, Reimer KJ, Poland JS. Extraction and speciation of arsenic in plants grown on arsenic contaminated soils. Talanta. 2007 Jun 15; 72(4): 1507-18.

[70] Aggett J, Kriegman MR. Preservation of arsenic (III) and arsenic (V) in samples of sediment interstitial water. Analyst. 1987; 112(2): 153-7.

[71] Segura M, Muñoz J, Madrid Y, Cámara C. Stability study of As (III), As (V), MMA and DMA by anion exchange chromatography and HG-AFS in wastewater samples. Analytical and bioanalytical chemistry. 2002 Oct; 374(3): 513-9.

[72] Mc Cleskey RB, Nordstrom DK, Meast AS. Peservation of water samples for arsenic (III/V) determinations: An evaluation of the literature and new analytical results. Applied Geochemistry. 2004; 9:995-1009.

[73] Hall GM. Stability of inorganic arsenic (III) and arsenic (V) in water samples. Journal of Analytical Atomic Spectrometry. 1999; 14(2): 205-13.

[74] Nikolaidis NP, Dobbs GM, Chen J, Lackovic JA. Arsenic mobility in contaminated lake sediments. Environmental pollution. 2004 Jun 1; 129(3): 479-87.

[75] Daus B, Weiss H, Mattusch J, Wennrich R. Preservation of arsenic species in water samples using phosphoric acid–Limitations and long-term stability. Talanta. 2006 Apr 15; 69(2): 430-4.

[76] Edwards M, Patel S, McNeill L, Chen HW, Frey M, Eaton AD, Antweiler RC, Taylor HE. Considerations in As analysis and speciation. Journal-American Water Works Association. 1998 Mar; 90(3): 103-13.

[77] Melanie G, Zagury JG, Deschenes L, Blouin J. Comparison of four extraction procedures to assess arsenate and arsenite species in contaminated soils. Environmental Pollution. 2009; 158:1890-1898.

[78] Van Herreweghe S, Swennen R, Vandecasteele C, Cappuyns V. Solid phase speciation of arsenic by sequential extraction in standard reference materials and industrially contaminated soil samples. Environmental pollution. 2003 Apr 1; 122(3): 323-42.

[79] Baig JA, Kazi TG, Shah AQ, Arain MB, Afridi HI, Kandhro GA, Khan S. Optimization of cloud point extraction and solid phase extraction methods for speciation of arsenic in natural water using multivariate technique. Analytica Chimica Acta. 2009 Sep 28; 651(1): 57-63.

[80] Wu J, Mester Z, Pawliszyn J. Speciation of organoarsenic compounds by polypyrrole-coated capillary in-tube

solid phase microextraction coupled with liquid chromatography/ electrospray ionization mass spectrometry. Analytica chimica acta. 2000 Dec 1; 424(2): 211-22.

[81] Pizarro I, Gómez M, Cámara C, Palacios MA. Arsenic speciation in environmental and biological samples: extraction and stability studies. Analytica Chimica Acta. 2003 Oct 24; 495(1-2): 85-98.

[82] Carabias-Martínez, R., Rodríguez-Gonzalo, E., Revilla-Ruiz, P. and Hernández-Méndez, J., 2005. Pressurized liquid extraction in the analysis of food and biological samples. *Journal of Chromatography A*, *1089*(1-2), pp. 1-17.

[83] Foster S, Maher W, Krikowa F, Apte S. A microwave-assisted sequential extraction of water and dilute acid soluble arsenic species from marine plant and animal tissues. Talanta. 2007 Feb 15; 71(2):537-49.

[84] Dietz C, Sanz J, Sanz E, Munoz-Olivas R, Cámara C. Current perspectives in analyte extraction strategies for tin and arsenic speciation. Journal of Chromatography A. 2007 Jun 15; 1153(1-2): 114-29.

[85] Chappell J, Chiswell B, Olszowy H. Speciation of arsenic in a contaminated soil by solvent extraction. Talanta. 1995 Mar 1; 42(3): 323-9.

[86] Georgiadis M, Cai Y, Solo-Gabriele HM. Extraction of arsenate and arsenite species from soils and sediments. Environmental Pollution. 2006 May 1; 141(1): 22-9.

[87] Garcia-Manyes S, Jiménez G, Padró A, Rubio R, Rauret G. Arsenic speciation in contaminated soils. Talanta. 2002 Aug 16; 58(1): 97-109.

[88] Yuan CG, He B, Gao EL, Lü JX, Jiang GB. Evaluation of extraction methods for arsenic speciation in polluted soil and rotten ore by HPLC-HG-AFS analysis. Microchimica Acta. 2007 Jun 1; 159(1-2): 175-82.

[89] Wu J, Zhang SD, Zhu XS. Study on the analytical method of inorganic arsenic species in environmental samples. Applied Ecology and Environmental Research. 2019 Jan 1; 17(4):7943-55.

[90] Rahman MS, Abdul Mazid Miah M, Khaled HM, Islam A, Panaullah GM. Arsenic concentrations in groundwater, soils, and irrigated rice in Southwestern Bangladesh. Communications in soil science and plant analysis. 2010 Aug 4; 41(16): 1889-95.

[91] Ruiz-Chancho MJ, Taleshi MS, Goessler W, Francesconi KA. A method for screening arsenolipids in fish oils by HPLC-ICPMS. Journal of Analytical Atomic Spectrometry. 2012; 27(3): 501-4.

[92] Gutiérrez-Ruiz ME, Ceniceros-Gómez AE, Villalobos M, Romero F, Santiago P. Natural arsenic attenuation via metal arsenate precipitation in soils contaminated with metallurgical wastes: II. Cumulative evidence and identification of minor processes. Applied geochemistry. 2012 Nov 1; 27(11): 2204-14.

[93] Tripathi P, Singh PC, Mishra A, Srivastava S, Chauhan R, Awasthi S, Mishra S, Dwivedi S, Tripathi P, Kalra A, Tripathi RD. Arsenic tolerant Trichoderma sp. reduces arsenic induced stress in chickpea (*Cicer arietinum*). Environmental Pollution. 2017 Apr 1; 223:137-45.

[94] Márquez-García B, Pérez-López R, Ruíz-Chancho MJ, López-Sánchez JF, Rubio R, Abreu MM, Nieto JM, Córdoba F. Arsenic speciation in soils and Erica andevalensis Cabezudo & Rivera and Erica australis L. from São Domingos Mine area, Portugal. Journal

of Geochemical Exploration. 2012 Aug 1; 119:51-9.

[95] Zhang X, Uroic MK, Xie WY, Zhu YG, Chen BD, McGrath SP, Feldmann J, Zhao FJ. Phytochelatins play a key role in arsenic accumulation and tolerance in the aquatic macrophyte *Wolffia globosa*. Environmental Pollution. 2012 Jun 1; 165:18-24.

[96] Yang F, Xie S, Wei C, Liu J, Zhang H, Chen T, Zhang J. Arsenic characteristics in the terrestrial environment in the vicinity of the Shimen realgar mine, China. Science of the Total Environment. 2018 Jun 1; 626:77-86.

[97] Bergqvist C, Herbert R, Persson I, Greger M. Plants influence on arsenic availability and speciation in the rhizosphere, roots and shoots of three different vegetables. Environmental pollution. 2014 Jan 1; 184:540-6.

[98] Rahman F, Chen Z, Naidu R. A comparative study of the extractability of arsenic species from silverbeet and amaranth vegetables. Environmental Geochemistry and Health. 2009 Apr; 31(1): 103-13.

[99] Paik MK, Kim MJ, Kim WI, Yoo JH, Park BJ, Im GJ, Park JE, Hong MK. Determination of arsenic species in polished rice using a methanol-water digestion method. Journal of the Korean Society for Applied Biological Chemistry. 2010 Oct; 53(5): 634-8.

[100] Kim JY, Kim WI, Kunhikrishnan A, Kang DW, Kim DH, Lee YJ, Kim YJ, Kim CT. Determination of arsenic species in rice grains using HPLC-ICP-MS. Food Science and Biotechnology. 2013 Dec 1; 22(6): 1509-13

[101] López-García I, Briceño M, Hernández-Córdoba M. Non-chromatographic screening procedure for arsenic speciation analysis in fish-based baby foods by using electrothermal atomic absorption spectrometry. Analytica chimica acta. 2011 Aug 5; 699(1): 11-7.

[102] Sun J, Yang Z, Lee H, Wang L. Simultaneous speciation and determination of arsenic, chromium and cadmium in water samples by high performance liquid chromatography with inductively coupled plasma mass spectrometry. Analytical Methods. 2015; 7(6): 2653-8.

[103] Pereira ÉR, Kopp JF, Raab A, Krupp EM, del Campo Menoyo J, Carasek E, Welz B, Feldmann J. Arsenic containing medium and long chain fatty acids in marine fish oil identified as degradation products using reversed-phase HPLC-ICP-MS/ESI-MS. Journal of Analytical Atomic Spectrometry. 2016; 31(9): 1836-45.

[104] Rubio R, Ruiz-Chancho MJ, López-Sánchez JF. Sample pre-treatment and extraction methods that are crucial to arsenic speciation in algae and aquatic plants. TrAC Trends in Analytical Chemistry. 2010 Jan 1; 29(1): 53-69.

[105] He B, Fang Y, Jiang G, Ni Z. Optimization of the extraction for the determination of arsenic species in plant materials by high-performance liquid chromatography coupled with hydride generation atomic fluorescence spectrometry. Spectrochimica Acta Part B: Atomic Spectroscopy. 2002 Nov 11; 57(11): 1705-11.

[106] Zheng J, Hintelmann H. HPLC-ICP-MS for a comparative study on the extraction approaches for arsenic speciation in terrestrial plant, *Ceratophyllum demersum*. Journal of Radioanalytical and Nuclear Chemistry. 2009 Apr; 280(1): 171-9.

[107] Mrak T, Šlejkovec Z, Jeran Z. Extraction of arsenic compounds from lichens. Talanta. 2006 Mar 15; 69(1): 251-8.

[108] Amaral CD, Nóbrega JA, Nogueira AR. Sample preparation for arsenic speciation in terrestrial plants—a review. Talanta. 2013 Oct 15; 115:291-9.

[109] Larios R, Fernández-Martínez R, LeHecho I, Rucandio I. A methodological approach to evaluate arsenic speciation and bioaccumulation in different plant species from two highly polluted mining areas. Science of the Total Environment. 2012 Jan 1; 414:600-7.

[110] Guzmán Mar JL, Hinojosa Reyes L, Mizanur Rahman GM, Kingston HS. Simultaneous extraction of arsenic and selenium species from rice products by microwave-assisted enzymatic extraction and analysis by ion chromatography-inductively coupled plasma-mass spectrometry. Journal of Agricultural and Food Chemistry. 2009 Apr 22; 57(8): 3005-13.

[111] Sadee BA, Foulkes ME, Hill SJ. An evaluation of extraction techniques for arsenic in staple diets (fish and rice) utilising both classical and enzymatic extraction methods. Food Additives & Contaminants: Part A. 2016 Mar 3; 33(3): 433-41.

[112] Mattusch J, Cimpean M, Wennrich R. Enzyme-assisted extraction of arsenic species from plant material. International Journal of Environmental and Analytical Chemistry. 2006 Aug 10; 86(9): 629-40.

[113] Raber G, Stock N, Hanel P, Murko M, Navratilova J, Francesconi KA. An improved HPLC–ICPMS method for determining inorganic arsenic in food: application to rice, wheat and tuna fish. Food chemistry. 2012 Sep 1; 134(1): 524-32.

[114] Heitkemper DT, Vela NP, Stewart KR, Westphal CS. Determination of total and speciated arsenic in rice by ion chromatography and inductively coupled plasma mass spectrometry. Journal of Analytical Atomic Spectrometry. 2001; 16(4): 299-306.

[115] Chen ML, Ma LY, Chen XW. New procedures for arsenic speciation: A review. Talanta. 2014 Jul 1; 125:78-86.

[116] Issa NB, Rajaković-Ognjanović VN, Marinković AD, Rajaković LV. Separation and determination of arsenic species in water by selective exchange and hybrid resins. Analytica chimica acta. 2011 Nov 7; 706(1): 191-8.

[117] Jia X, Gong D, Wang J, Huang F, Duan T, Zhang X. Arsenic speciation in environmental waters by a new specific phosphine modified polymer microsphere preconcentration and HPLC–ICP-MS determination. Talanta. 2016 Nov 1; 160:437-43.

[118] Huang C, Xie W, Li X, Zhang J. Speciation of inorganic arsenic in environmental waters using magnetic solid phase extraction and preconcentration followed by ICP-MS. Microchimica Acta. 2011 Apr 1; 173(1-2): 165-72.

[119] An MI, Zhang X, Yang T, Chen M, Wang J. Uptake and speciation of inorganic arsenic with cellulose fibre coated with yttrium hydroxide layer as a novel green sorbent. Chinese Journal of Chemistry. 2012 Sep; 30(9): 2225-31.

[120] Qian S, Huang Z, Fu J, Kuang J, Hu C. Preconcentration of ultra-trace arsenic with nanometre-sized TiO 2 colloid and determination by AFS with slurry sampling. Analytical Methods. 2010; 2(8): 1140-3.

[121] Chen M, Lin Y, Gu C, Wang J. Arsenic sorption and speciation with branch-polyethyleneimine modified carbon nanotubes with detection by atomic fluorescence spectrometry. Talanta. 2013 Jan 30; 104:53-7.

[122] Güell R, Fontàs C, Salvadó V, Anticó E. Modelling of liquid–liquid extraction and liquid membrane separation of arsenic species in environmental matrices. Separation and purification technology. 2010 May 11; 72(3): 319-25.

[123] Majidi B, Shemirani F. In situ solvent formation microextraction in the presence of ionic liquid for preconcentration and speciation of arsenic in saline samples and total arsenic in biological samples by electrothermal atomic absorption spectrometry. Biological trace element research. 2011 Oct; 143(1): 579-90.

[124] Shraim A, Chiswell B, Olszowy H. Speciation of arsenic by hydride generation–atomic absorption spectrometry (HG–AAS) in hydrochloric acid reaction medium. Talanta. 1999 Dec 1; 50(5): 1109-27.

[125] Anawar HM. Arsenic speciation in environmental samples by hydride generation and electrothermal atomic absorption spectrometry. Talanta. 2012 Jan 15; 88:30-42.

[126] Cathum SJ, Brown CE, Obenauf A, Punt M. Speciation of arsenic using chelation solvent extraction and high performance liquid chromatography. CLEAN–Soil, Air, Water. 2007 Feb; 35(1): 71-80.

[127] Akter KF, Chen Z, Smith L, Davey D, Naidu R. Speciation of arsenic in ground water samples: A comparative study of CE-UV, HG-AAS and LC-ICP-MS. Talanta. 2005 Dec 15; 68(2): 406-15.

[128] Xie D, Mattusch J, Wennrich R. Separation of Organoarsenicals by Means of Zwitterionic Hydrophilic Interaction Chromatography (ZIC®-HILIC) and Parallel ICP-MS/ESI-MS Detection. Engineering in Life Sciences. 2008 Dec;8(6): 582-8.

[129] Shen S, Li XF, Cullen WR, Weinfeld M, Le XC. Arsenic binding to proteins. Chemical reviews. 2013 Oct 9; 113(10): 7769-92.

[130] Migoni D, Papadia P, Cannito F, Fanizzi FP. Sequential Extraction Analysis of Arsenic in Soil Samples Collected in an Agricultural Area of Brindisi, Apulia (Italy), in the Proximity of a Coal-Burning Power Plant. Applied Sciences. 2021 Jan; 11(5): 2115.

[131] Watts MJ, O'reilly J, Marcilla AL, Shaw RA, Ward NI. Field based speciation of arsenic in UK and Argentinean water samples. Environmental geochemistry and health. 2010 Dec 1; 32(6): 479-90.

[132] Sadee BA, Foulkes ME, Hill SJ. A study of arsenic speciation in soil, irrigation water and plant tissue: a case study of the broad bean plant, *Vicia faba*. Food chemistry. 2016 Nov 1; 210:362-70.

[133] Sugár É, Tatár E, Záray G, Mihucz VG. Field separation-based speciation analysis of inorganic arsenic in public well water in Hungary. Microchemical Journal. 2013 Mar 1; 107:131-5.

[134] Hong S, Choi SD, Khim JS. Arsenic speciation in environmental multimedia samples from the Youngsan River Estuary, Korea: A comparison between freshwater and saltwater. Environmental Pollution. 2018 Jun 1; 237:842-50.

[135] Nogueira R, Melo EA, Figueiredo JL, Santos JJ, Nascimento Neto AP. Arsenic Speciation in Fish and Rice by HPLC-ICP-MS Using Salt Gradient Elution. Journal of the Brazilian Chemical Society. 2018 Aug; 29(8): 1593-600.

[136] Rahman MM, Chen Z, Naidu R. Extraction of arsenic species in soils using microwave-assisted extraction detected by ion chromatography

coupled to inductively coupled plasma mass spectrometry. Environmental geochemistry and health, 2009; 31(1): 93-102.

[137] Casiot C, Donard OF, Potin-Gautier M. Optimization of the hyphenation between capillary zone electrophoresis and inductively coupled plasma mass spectrometry for the measurement of As-, Sb-, Se-and Te-species, applicable to soil extracts. Spectrochimica Acta Part B: Atomic Spectroscopy. 2002 Jan 9; 57(1): 173-87

[138] Qu H, Mudalige TK, Linder SW. Arsenic speciation in rice by capillary electrophoresis/inductively coupled plasma mass spectrometry: enzyme-assisted water-phase microwave digestion. Journal of agricultural and food chemistry. 2015 Apr 1; 63(12): 3153-60.

[139] Richardson DD, Kannamkumarath SS, Wuilloud RG, Caruso JA. Hydride generation interface for speciation analysis coupling capillary electrophoresis to inductively coupled plasma mass spectrometry. Analytical chemistry. 2004 Dec 1; 76(23): 7137-42.

[140] Duester L, Diaz-Bone RA, Kösters J, Hirner AV. Methylated arsenic, antimony and tin species in soils. Journal of Environmental Monitoring. 2005; 7(12): 1186-93.

[141] Arroyo-Abad U, Mattusch J, Möder M, Elizalde-González MP, Matysik FM. Identification of degradation products of phenylarsonic acid and o-arsanilic acid in contact with suspensions of soils of volcanic origin. Talanta. 2012 Sep 15; 99:310-5.

[142] Arroyo-Abad U, Mattusch J, Möder M, Elizalde-González MP, Wennrich R, Matysik FM. Identification of roxarsone metabolites produced in the system: Soil–chlorinated water–light by using HPLC-ICP-MS/ESI-MS, HPLC-ESI-MS/MS and High Resolution Mass Spectrometry (ESI-TOF-MS). Journal of Analytical Atomic Spectrometry. 2011; 26(1):171-7.

[143] Dai J, Chen C, Gao AX, Tang Z, Kopittke PM, Zhao FJ, Wang P. Dynamics of Dimethylated Monothioarsenate (DMMTA) in Paddy Soils and Its Accumulation in Rice Grains. Environmental Science & Technology. 2021 Jun 10.

[144] Huang K, Peng H, Gao F, Liu Q, Lu X, Shen Q, Le XC, Zhao FJ. Biotransformation of arsenic-containing roxarsone by an aerobic soil bacterium Enterobacter sp. CZ-1. Environmental Pollution. 2019 Apr 1; 247:482-7.

[145] Yu X, Xiong C, Jensen KB, Glabonjat RA, Stiboller M, Raber G, Francesconi KA. Mono-acyl arsenosugar phospholipids in the edible brown alga Kombu (*Saccharina japonica*). Food Chemistry. 2018 Feb 1; 240:817-21.

[146] Vriens B, Ammann AA, Hagendorfer H, Lenz M, Berg M, Winkel LH. Quantification of methylated selenium, sulfur, and arsenic in the environment. PlOS one. 2014 Jul 21; 9(7): 102906.

[147] Bluemlein an CG, Jiang GB, He B. Evaluation of the extraction methods for arsenic speciation in rice straw, *Oryza sativa* L., and analysis by HPLC-HG-AFS. Journal of Analytical Atomic Spectrometry. 2005; 20(2): 103-10.

[148] de Oliveira RM, Antunes AC, Vieira MA, Medina AL, Ribeiro AS. Evaluation of sample preparation methods for the determination of As, Cd, Pb, and Se in rice samples by GF AAS. Microchemical Journal. 2016 Jan 1; 124:402-9.

[149] Yuan CG, Jiang GB, He B. Evaluation of the extraction methods for arsenic speciation in rice straw, *Oryza sativa* L., and analysis by

HPLC-HG-AFS. Journal of Analytical Atomic Spectrometry. 2005; 20(2): 103-10.

[150] Huynh T, Harris HH, Zhang H, Noller BN. Measurement of labile arsenic speciation in water and soil using diffusive gradients in thin films (DGT) and X-ray absorption near edge spectroscopy (XANES). Environmental Chemistry. 2015 Feb 17; 12(2): 102-11.

[151] Voegelin A, Weber FA, Kretzschmar R. Distribution and speciation of arsenic around roots in a contaminated riparian floodplain soil: Micro-XRF element mapping and EXAFS spectroscopy. Geochimica et Cosmochimica Acta. 2007 Dec 1; 71(23): 5804-20.

[152] Wang D, Root RA, Chorover J. Biochar-templated surface precipitation and inner-sphere complexation effectively removes arsenic from acid mine drainage. Environmental Science and Pollution Research. 2021 Apr 18:1-5.

[153] Frau F, Rossi A, Ardau C, Biddau R, Da Pelo S, Atzei D, Licheri C, Cannas C, Capitani G. Determination of arsenic speciation in complex environmental samples by the combined use of TEM and XPS. Microchimica Acta. 2005 Nov; 151(3): 189-201.

[154] Bullen JC, Torres-Huerta A, Salaün P, Watson JS, Majumdar S, Vilar R, Weiss DJ. Portable and rapid arsenic speciation in synthetic and natural waters by an As (V)-selective chemisorbent, validated against anodic stripping voltammetry. Water research. 2020 May 15; 175:115650.

[155] Giuffrè O, Aiello D, Chillè D, Napoli A, Foti C. Binding ability of arsenate towards Cu 2+ and Zn 2+: thermodynamic behavior and simulation under natural water conditions. Environmental Science: Processes & Impacts. 2020; 22(8): 1731-42.

[156] Jackson BP, Liba A, Nelson J. Advantages of reaction cell ICP-MS on doubly charged interferences for arsenic and selenium analysis in foods. Journal of analytical atomic spectrometry. 2015; 30(5): 1179-83.

[157] Stucker VK, Silverman DR, Williams KH, Sharp JO, Ranville JF. Thioarsenic species associated with increased arsenic release during biostimulated subsurface sulfate reduction. Environmental science & technology. 2014 Nov 18; 48(22): 13367-75.

[158] Sele V, Sloth JJ, Holmelid B, Valdersnes S, Skov K, Amlund H. Arsenic-containing fatty acids and hydrocarbons in marine oils–determination using reversed-phase HPLC–ICP-MS and HPLC–qTOF-MS. Talanta. 2014 Apr 1; 121:89-96.

[159] Khan M, Francesconi KA. Preliminary studies on the stability of arsenolipids: Implications for sample handling and analysis. Journal of Environmental Sciences. 2016 Nov 1; 49:97-103.

[160] Maher WA, Foster S, Krikowa F, Duncan E, St John A, Hug K, Moreau JW. Thio arsenic species measurements in marine organisms and geothermal waters. Microchemical Journal. 2013 Jul 1; 111:82-90.

[161] Lai VW, Kanaki K, Pergantis SA, Cullen WR, Reimer KJ. Arsenic speciation in freshwater snails and its life cycle variation. Journal of Environmental Monitoring. 2012; 14(3): 743-51.

[162] Raab A, Newcombe C, Pitton D, Ebel R, Feldmann J. Comprehensive analysis of lipophilic arsenic species in a brown alga (*Saccharina latissima*). Analytical chemistry. 2013 Mar 5; 85(5): 2817-24.

[163] Feldmann J, Raab A, Krupp EM. Importance of ICPMS for speciation

analysis is changing: future trends for targeted and non-targeted element speciation analysis. Analytical and bioanalytical chemistry. 2018 Jan;410(3):661-7.

[164] Schmeisser E, Goessler W, Kienzl N, Francesconi KA. Volatile analytes formed from arsenosugars: determination by HPLC–HG-ICPMS and implications for arsenic speciation analyses. Analytical chemistry. 2004 Jan 15; 76(2): 418-23.

[165] Yang H, He M. Distribution and speciation of selenium, antimony, and arsenic in soils and sediments around the area of Xikuangshan (China). CLEAN–Soil, Air, Water. 2016 Nov; 44(11): 1538-46.

[166] Kristan U, Kanduč T, Osterc A, Šlejkovec Z, Ramšak A, Stibilj V. Assessment of pollution level using *Mytilus galloprovincialis* as a bioindicator species: The case of the Gulf of Trieste. Marine pollution bulletin. 2014 Dec 15; 89(1-2):455-63.

[167] Rivera N, Hesterberg D, Kaur N, Duckworth OW. Chemical speciation of potentially toxic trace metals in coal fly ash associated with the Kingston fly ash spill. Energy & Fuels. 2017 Sep 21; 31(9): 9652-9.

[168] Cui JL, Zhao YP, Li JS, Beiyuan JZ, Tsang DC, Poon CS, Chan TS, Wang WX, Li XD. Speciation, mobilization, and bioaccessibility of arsenic in geogenic soil profile from Hong Kong. Environmental pollution. 2018 Jan 1; 232:375-84.

[169] Whaley-Martin KJ, Koch I, Reimer KJ. Determination of arsenic species in edible periwinkles (*Littorina littorea*) by HPLC–ICPMS and XAS along a contamination gradient. Science of the total environment. 2013 Jul 1; 456:148-53.

[170] Madsen AD, Goessler W, Pedersen SN, Francesconi KA. Characterization of an algal extract by HPLC-ICP-MS and LC-electrospray MS for use in arsenosugar speciation studies. Journal of Analytical Atomic Spectrometry. 2000; 15(6): 657-62.

[171] Maher WA, Ellwood MJ, Krikowa F, Raber G, Foster S. Measurement of arsenic species in environmental, biological fluids and food samples by HPLC-ICPMS and HPLC-HG-AFS. Journal of Analytical Atomic Spectrometry. 2015; 30(10): 2129-83.

Section 2

Arsenic Removal and Remediation

Chapter 3

Removal of Arsenic - "A Silent Killer" in the Environment by Adsorption Methods

Ashok Kumar, Kaman Singh, Utkarsh Dixit, Rayees Ahmad Bhat and Satya Prakash Gupta

Abstract

Water is one of the most essential requirements for living being to survive because 70–80% of the mass of most living bodies consists of water and various mineral and organic salts. Water is also most important component of our environment. Large amount of water is used in various industries or commercial level or domestic level and finally effluent water is loaded with large amount of pollutants such as organic chemicals (surfactants, dyes, phenols etc.), inorganic hazardous heavy metals (As in present case) microbes (bacteria, fungi etc.) pollutants particulate etc. Arsenic is a natural metalloid chemical that may be present in groundwater and surface water gets polluted, hence, aquatic life of plants and animals is disturbed and cause abnormal growth and various diseases, hence, short term or long term changes occurs in ecosystem. Hence, treatment of wastewater is essentially required before discharge effluent wastewater into ponds or lagoons, drains and rivers. Arsenic is one such element that contaminates the environment as reported in several countries. The largest population at risk is in Bangladesh followed by India (West Bengal). Arsenic is familiar as silent killer because dissolved in water, it is colorless, odorless, and tasteless, yet consumption of relatively small doses of this element in its most toxic forms can cause rapid and violent death. It is a human carcinogen in water over a wide range of pH values, having harmful effects on both human health and environment, even at low concentration. Because of this effect, the World Health Organization (WHO) and the US Environmental Protection Agency (USEPA) set the arsenic standard for drinking water at .010 ppm to protect consumers served by public water systems. Ingestion only poses health problems if a dangerous amount of arsenic enters the body. Then, it can lead to cancer, liver disease, coma, and death. There is no effective treatment for arsenic toxicity. Only the removal of arsenic from aqueous system can prevent the toxicity. A great deal of research over recent decades has been done to lower the concentration of arsenic in drinking water and still there is a need to develop ecofriendly techniques. Existing major arsenic removal technologies include oxidation, adsorption, precipitation, coagulation and membrane separation. This book chapter presents a systematic description of current status of research in the area of arsenic removal from contaminated water and comparison of all technologies available with more emphasis on adsorption.

Keywords: Arsenic, Adsorption, Environmental and Health Aspects

1. Introduction

Water is one of the most essential requirements for living being to survive because all physiochemical processes of body require aqueous medium this is due to Moreover, 70–80% of the mass of most living bodies consists of water and various mineral and organic salts [1]. Large amount of water is used in various industries or commercial level or domestic level and finally effluent water is loaded with large amount of pollutants such as organic chemicals (surfactants, dyes, phenols, etc.), inorganic hazardous heavy metals (As, Hg, Cd, Pb, etc.) microbes (bacteria, fungi, etc.) pollutants particulate etc. Arsenic is a natural metalloid chemical that may be present in groundwater and surface water gets polluted, hence, aquatic life of plants and animals is disturbed and cause abnormal growth and various diseases, hence, short term or long term changes occurs in ecosystem [2–4]. Very low concentration (1.50 mg/L) of surfactant is lethal for microorganism, even about 0.50 mg/L is harmful for aquatic life. For human life limit of anionic surfactant concentration should up to 1.0 mg/L. High concentration of surfactant in drinking water causes cancer, irritation, dermatitis, eyes disorder. Water has a broad impact on all aspects of human life including but not limited to health, food, energy, and economy. In addition to the environmental, economic, and social impacts of poor water supply and sanitation, the supply of fresh water is essential for the safety of children and the poor. It is estimated that 10.0–20.0 million people die every year due to water-borne and nonfatal infection causes death of more than 200.0 million people every year. Every day, about 5,000.0–6,000.0 children die due to the water-related problem of diarrhea. There are currently more than 0.78 billion people around the world who do not have access to safe water resources resulting in major health problems [5]. Hence, it increases the permeability of cell membrane, and hence, removal of these pollutants are necessary from industrial and household water [6].

Release of hazardous pollutants and their dispersal in the environment can cause adverse impacts on the environment and to public health [7]. These pollutants are more easily controlled when they are generated than after they are dispersed. It is therefore of prime necessity to design treatment processes that isolate and remove the contaminants at their source [8]. Various methods have been developed in the past decades for treatment of waste water for arsenic as biological, physical and chemical methods. In these methods includes chemical precipitation, flocculation-coagulation [9], electro-flotation, electrochemical destruction, electrochemical coagulation [10], biological degradation [11], ozonation, hydrogen peroxide [12], reverse osmosis [13] etc. All these methods are good but have limited applications and all are very expensive, hence, could not use by small industry for wastewater treatments. For that reason there is need for some conventional method which are economically and ecofriendly for wastewater treatments (As in present case). Adsorption is one of the most extensively applied techniques for the removal of pollutants from the industrial effluents. The prominent and emerging trend of subjecting biosorbents in the adsorption technology is mainly because of their natural existence, abundance, renewable, biodegradable and economic features. The adsorption isotherm equations used to describe the experimental data and the thermodynamic assumptions of the various models [14–17].

2. Distribution of arsenic in various parts of world water bodies

In the nature arsenic element is found in soil, water and sediments as arsenic oxides. The common chemical oxidation of arsenic in the nature are −3, +3, 0 and + 5.¨, etc. According to the national agency for the research of cancer

Water type	As contamination (μg/Liter)	Ref.
Ground water in various part of world	1 -73.6	[19]
US Rivers	0.1-2	[20]
Gomti River, India	1.29-9.62	[21]
Manchar Lake, Pakistan	60.45	[22]
Alpine River, France	0.1-263	[23]
Zenne River, Belgium	0.1-263	[24]

Table 1.
The distribution of arsenic in natural water can be given as below.

arsenic compounds are put in group 1 carcinogens [18]. According to World Health Organization the maximum contaminant level of arsenic in the drinking water should not be greater than 10 μg/liter. By the dissolution of minerals, microbial activity arsenic can release into the aquatic environment. Living organism can be affected by arsenic through drinking water. Arsenic polluted water causes damage to central nervous system, kidney, liver, lungs and skin in humans. Further chronic arsenic can cause cardio diseases, hypertension and affects vascular system. Long use of arsenic contaminated water can also cause pigmentation of skin, development of hard paths on the palm of humans. Therefore removal of arsenic from waste water has been remained a subject of concern (**Table 1**) [17, 18].

3. Arsenic removal by adsorption

Many methods have been given for Arsenic removal which include chemical precipitation [25], adsorption [26], ion exchange [27], reverse osmosis [28], and electro-dialysis [29]. Out of all these methods adsorption method found to be cheap and best for the removal of arsenic. There are many kinds of adsorbents available for arsenic removal, for their convenient study we can divide them in two categories. The first category is metal and its alloys such as manganese oxide, activated alumina and iron compounds; the second category is activated carbon obtained from red mud, coconut shell and other carbon-like materials. Larger is the surface area of the adsorbent, stronger is its adsorption effect. It is found that that adsorption of As(V) is better than As(III) (**Figure 1**).

3.1 Adsorption isotherm models

3.2 The selection of isotherm models

The Akaike information criterion is used for the selection of appropriate model under the situation of when data fitted by more than one model. The Akaike information criterion is represented as below [30].

$$AlC = 2k - 2\ln(L)$$

Where k is the number of parameters in the model and L is the maximum value of the likelihood function for the model.

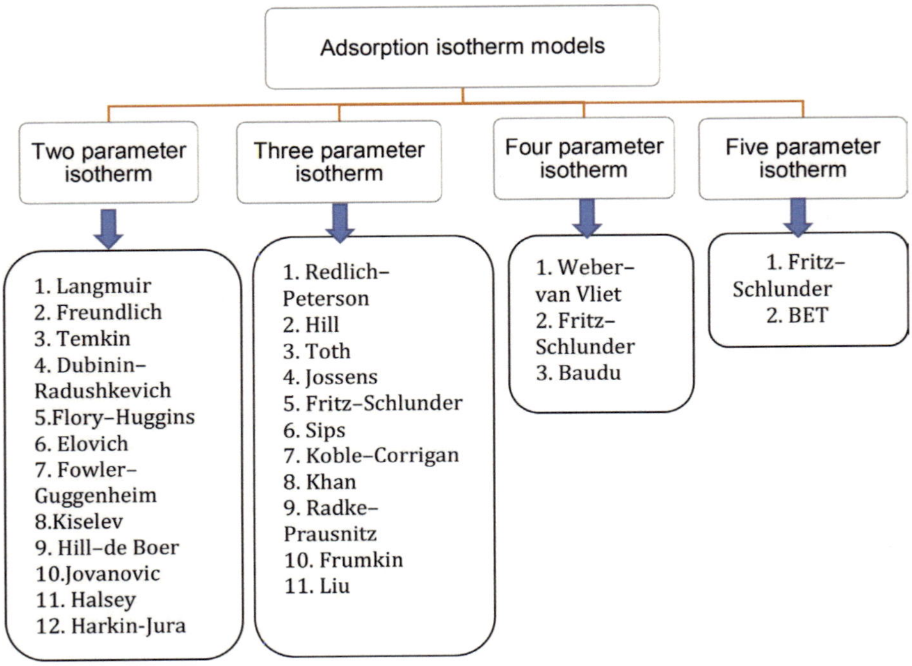

Figure 1.
Different adsorption isotherm models used to describe the experimental data and the thermodynamic assumptions of adsorption method.

3.3 Arsenic removal by iron based adsorbents

Iron based adsorbents are extensively developed and used for the removal of arsenic from water [31] (**Table 2**). Some adsorbents such as granular ferric hydroxide (GFH) and zero-valent iron have been produced on an industrial scale as

Adsorbents	Surface area (m^2 g^{-1})	Initial concentration (mg L^{-1})	pH	Adsorption capacity (mg g^{-1}) As(III)	Adsorption capacity (mg g^{-1}) As (V)	Ref.
Granular ferric hydroxide (GFH)	240–300	As(V):0.1	6.5-7.5	—	1.1	[32]
α-FeOOH nanoparticles	167.8	As(V):100	3.0	—	76	[33]
Ultrafine Fe2O3 nanoparticles	162	—	4.0	47	95	[34]
Ultrafine FeOOH	135	As(V): 20	7.0	—	37.3	[35]
β -FeOOH nanoparticles	330	As(V): 20	7.5	—	120	[36]
Magnetite–magnetite nanoparticles	49	As(III): 1.5 As(V): 1.5	2.0	3.69	3.71	[37]
γ -Fe2O3 nanoparticles	41-49	As(V): 1	7.0	—	2.9	[38]

Table 2.
Different iron compounds used as adsorbent for As(III) and As(V) adsorption from aqueous system.

commercial adsorbents [39]. Iron oxy-hydroxides are mostly used as the adsorbent because of their easy accessibility. The commonly used iron oxy-hydroxides such as, akaganeite (β-FeOOH), goethite (α-FeOOH), lepidocrocite (γ-FeOOH), ferrihydrites ($Fe_{10}O_{14}(OH)_2$), green rusts can be chemically synthesized by the precipitation of Fe(III) or Fe(+2) salts through the hydrolysis and oxidation processes [40].

3.4 Adsorption of arsenic using activated carbon (AC)

Activated carbon is the most effective and efficient and most widely used because of its versatile nature and convenient for removal of As(III) and As(V). The activated carbon is micro-porous form of the carbon having large surface area and greater number of pores [41]. Arsenic metal cannot be removed completely by simple chemical or physical treatments. So, in this case, AC is used to remove the toxic metals completely from aqueous solution because there is formation of surface complexes between acidic surface functional groups of AC and the metal ions [42]. The removal efficiency is dependent on various factors, like solution concentration initial pH, ionic strength, adsorbent modification methods, nature of adsorbate, chemical nature of AC, and physical properties (surface area, porosity) [43–45]. Adsorption of both arsenic is largely affected by environmental factors such as pH solution, ionic strength, and coexisting substances such as anions, cations, and organic matter. The adsorption capacity of the adsorbents depends not only on the surface area, pore volume, and particles size but also on a combination of all factors, surface chemistry, and pore structure. Specific area of the adsorbents does not contribute to the adsorption capacity on the removal of arsenic from water. Therefore selection of the adsorbents for removal of arsenic should be based on a combination of all factors for the adsorbents and adsorbate.

3.4.1 Preparation of adsorbents

There are two different methods for preparation of activated carbon one is physical and other one chemical. Chemical activation is better than physical because it requires lower temperature and global yield is more since burn-off does not require (**Figure 2**) [45].

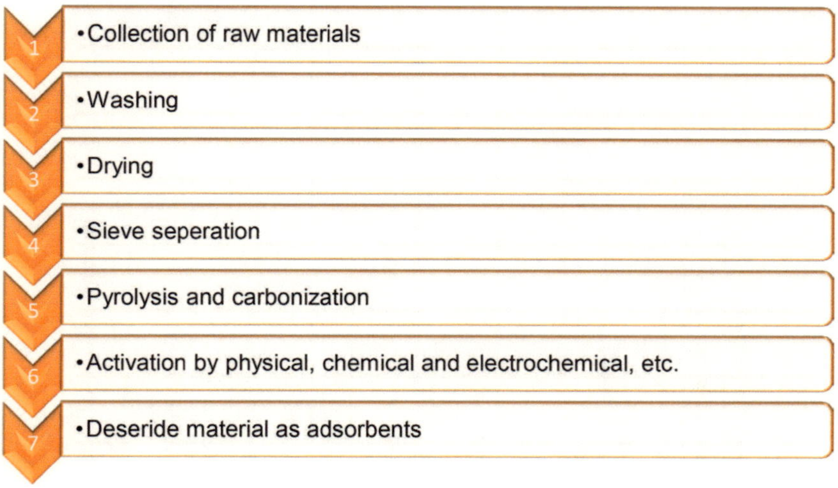

Figure 2.
General flow diagram for the preparation of adsorbents.

Adsorbents	Surface area (m^2 g^{-1})	Initial concentration (mg/L)	pH	Capacity As (III) mg/gm	Capacity As (V) mg/gm	Isotherm model fitted	Ref.
Coconut husk	206	50-60	12	146.30	—	—	[46]
Coconut shell with 3% ash	1150-1250	0-200	6	—	2.4	Langmuir	[47]
Ce-Ti oxide adsorbent	137.4	50	6.5	7.5	—	Langmuir	[48]
Fe_3O_4 loaded activated carbon	349	200	8	—	204.2	Freundlich	[49]
Char Carbon	36.48	157-992	2-3	89	34.46	—	[50]
Red mud	130	7.5-220	4.5	0.541	7.642	—	[51]
Pine Leaves	—	10	4	—	3.27	Both Langmuir and Flory-Huggins	[52]
Oat Hulls	520	25-200	5	—	3.09	Langmuir	[53]
Iron Impregnated GAC	650	0-5	7	—	1.95	Langmuir	[54]
AC (apricot stone)	1547	4.5	—	—	0.034	Both Freundlich and Dubinin–Radushkevich	[55]
Fe (II) loaded	1231	4.5	3	—	2.023	Redlich–Peterson	[56]
Fe (III) loaded	987	4.5	3	—	3.009	Redlich–Peterson	[57]
Sugar beet pulp	821	0-5	9.1-9.4	—	0.691	Langmuir	[58]
Sugar beet pulp-Fe	762	0-3.5	8.5-9.1	—	2.9	Langmuir	[59]
Empty fruit bunch biochar	1890	3-300	9.4	18.9	5.5	Langmuir	[60]
Rice husk biochar	25.16	3-300	8.5	19.3	7.1	Langmuir	[61]
Fe coated empty fruit bunch biochar	—	3-300	9.4	31.4	15.2	Both Langmuir and Freundlich	[62]
Fe coated rice husk biochar	—	3-300	8.5	30.7	16	Langmuir	[63]
Leonardite char	65.68	1-80	7	4.46	8.40	Langmuir	[64]
Tea Waste	—	1-100	7	189	154	Langmuir	[65]

Adsorbents	Surface area (m^2 g^{-1})	Initial concentration (mg/L)	pH	Capacity As (III) mg/gm	Capacity As (V) mg/gm	Isotherm model fitted	Ref.
Concrete Sludge	23	10-700	7	—	175	Langmuir	[66]
Chitosan	3.1	0.025-2	5.6-6.2	—	0.73	Langmuir	[67]
Iron- Chitosan flakes	1.44	1-10	7	16.2	22.5	Langmuir	[68]
Iron- Chitosan granules	96.8	1-10	7	2.32	2.24	Langmuir	[69]

Table 3.
Different adsorbents and their adsorption capacity for the effective removal As(III) and As(V) employing various adsorption isotherm models.

4. Adsorbents comparison

The adsorption capacity of various adsorbents for the effective removal arsenic employing various adsorption isotherm models have been summarized in **Table 3.**

5. Characteristics of various methods for the removal of arsenic from aqueous system

In order to improve the current analytical methods by removing or treating arsenic from aqueous system, a number of methods have been developed and reviewed in this chapter, which included several method with their characteristics properties (**Table 4**).

Such information should be taken into consideration in order to improve the current methods or develop new advanced methods. The best analytical methods for aresenic speciation are considered those, including chromatographic separations based on adsorption coupled with a sensitive detection system. Specific sorbents and exchange resins have been developed and applied recently for this purpose. Apart from the chromatographic and non-chromatographic methods for the arsenic

Method	Method in detail	Characteristics	Ref.
Physical	Mixing both contaminated and uncontaminated soils	High cost/usage to smaller-scale operations	[70–72]
	Washed with sulfuric acid, nitric acid, phosphoric acid, and hydrogen bromide	Chemicals usage/high cost/usage to smaller-scale operations	
	Immobilize soluble arsenates using cement	Successfully used to stabilize As-rich	
	Emphasis on stabilization/ solidification (S/S)	Treating As containing wastes in water	
	Soil flushing using aqueous solutions using surfactants and solvents Adsorption by using specific media,	Applied in the field, efficiency can vary from 0% to almost 100%	
Chemical	immobilization, modified coagulation along with filtration, precipitations, immobilizations, and complexation reactions	Economic but often displayed lower Efficiencies (<90%)	[70, 73–76]
	Formation of stable phases, for example, insoluble $FeAsO_4$ (and hydrous species of this compound such as scorodite, $FeAsO_4.2H_2O$)	Use of selective stabilizing amendments is a challenging task	
	Stabilization method using nanosized oxides and Fe(0) (particle size of 1 to 100 nm)	Gained popularity/ high success rate, but it could be expensive when remediating a large area	
Intrinsic bioremediation	Degradation of arsenic by naturally occurring microorganism	More suitable for remediation of soil with a low level of contaminants	[77]
Engineered bioremediation	Optimizing the environment condition to promote the proliferation and activity of microorganisms	Favorable method used in high contaminated area	[78]

Method	Method in detail	Characteristics	Ref.
Microbial oxidation	Immobilization of As in the solid phase	Required biological activity, and microbiological molecular analysis/ involved adsorption or co-precipitation with Fe-ox hydroxides	[79]
Physiochemical	Filtration or coagulation sedimentation, osmosis or electro dialysis, adsorptions, and chemical precipitations	Widely accepted in some places	[70, 73–76]
Biological	Such as phytoremediation by using aquatic plants or microbial detoxification of arsenic	Widely accepted in some places	[70, 73–76]
Phytoremediation	Using "free-floating plants such as water hyacinth" Using aquatic rooted plants such as Agrostis sp., *Pteris vittata*, and *Pteris cretica*	Widely accepted in some places	[70, 73–76]
Methylations	Biomethylations (by As(III) (by As(III) S-adenosylmethionine methyltransferase	Is a reliable biological process of removing arsenic from aquatic mediums	[70]
Reduction	Reduction of arsenate into arsenite by microorganisms via dissimilatory reduction mechanism	Should be carried out in facultative anaerobe or strict anaerobe condition	[70]
Oxidation	Using heterotrophic bacteria and chemoautotrophic bacteria to oxidize arsenite into a less toxic arsenate	Should be carried out in controlled environment	[80–83]

Table 4.
Characteristics of various methods for the removal of arsenic from aqueous system.

species separation, simple and cost-effective electrochemical methods were developed recently based on the distinct As-species electrochemical properties [84].

6. Conclusion

This chapter on removal of arsenic - "a silent killer" in the environment by adsorption methods has been discussed. Moreover, recently the consumers have become very much conscious about the environment, renaissance of eco-friendly products and process, which has thus become also important now. Thus, revival of natural arsenic application on various areas and summary of earlier researches on standardization of its method of extraction, mordanting, process variables and even natural finishing, etc. have been elaborated in this chapter. Thus this part has become a unique comprehensive chapter for information on removal of 'arsenic' a silent killer in the environment by adsorption method. A brief review of the removal of arsenic ions from water using iron-based adsorbents has been presented. A few adsorbents discussed in this chapter include relative advantages and disadvantages of adsorbents used for the removal of arsenic from water have been mentioned. The mechanism of arsenic adsorption on iron-containing adsorbents was summarized. Overall, there exist significant progress and benefit on using adsorption process for removing arsenic species from groundwater in a practical way to make potable water accessible for the rural population.

Conflict of interest

There is no conflict of interest.

Author details

Ashok Kumar, Kaman Singh*, Utkarsh Dixit, Rayees Ahmad Bhat and Satya Prakash Gupta
Surface Science Laboratory, Department of Chemistry, Babasaheb Bhimrao Ambedkar University (A Central University), Lucknow, Uttar Pradesh, India

*Address all correspondence to: drkamansignh@yahoo.com

References

[1] Zhegunov G., The Dual Nature of Life, The Frontiers Collection, Springer-Verlag Berlin Heidelberg, 1(300), 2012.

[2] Azarmi R. and Ashjaran A. "Type and application of some common surfactants" Journal of Chemical and Pharmaceutical Research, 7(2):632-640, 2015.

[3] Elsberry C.C., Belanger S. etal "Environmental Safety of the Use of Major Surfactant Classes in North America", Critical Reviews in Environmental Science and Technology, 44(17): 2014.

[4] Purakayastha P.D., etal "Adsorbent selection for anionic surfactant removal from water", Indian Journal of Chemical Technology, 12(281-284), 2005.

[5] Bagul V. R.; Shinde D. N,etal "Causes and Impacts of Water Pollution on Rivers in Maharashtra-A Review" , Research Journal of Chemical and Environmental Sciences, 3 [6] 2015.

[6] Tripathy D. B., Mishra A. eatal "Biodegradability of Laundry Detergent Surfactants", Journal of Surfactants and Detergents, 5(130-136), 2015.

[7] Mathur N., Bhatnagar P., "Review of the Mutagenicity of Textile Dye Products" Universal Journal of Environmental Research and Technology, 2(1-18), 2012.

[8] ElsalamM.R.M.A. "Studies on the Use of Surface Active Agents for the Removal of Some Pollutants from Dilute Aqueous", Hot Labs. Center- Atomic Energy, 2007.

[9] Bittton G., WileyJ. and sons "Wastewater microbiology" 3 (63-65),1994. DOI: 10.1002/978047974704.ch5

[10] Ratana, Padhi B.s.Polltution due to synthetic dyes toxicity and carcinogenicity studies and remediation", International Jouranal of Enviornmental Sciences, 3(940-955), 2012.

[11] Mazumdar D., "Process evaluation and treatability study of wastewater in textile dyeing industry", International Journal of Energy and Environment, 2(1053-1066), 2011.

[12] Thakur S., Chauhan M.S., "Treatment of waste water by electro coagulation" 5(104-110), 2016.

[13] Lavanya C., Dhankar R. etal "Degradation of toxic dyes: A Review", International Journal of Current Microbiology and Applied Sciences, 3(186-199), 2014.

[14] Rangabhashiyam S., N. Anu, Nandagopal M.S.G., Selvaraju N., "Relevance of isotherm models in biosorption of pollutants by agricultural byproducts.", Journal of Environmental Chemical Engineering, 2(398-414), 2014.

[15] C. G. Lee, P. J. J. Alvarez, A. Nam, et al., Effective removal of arsenate from wastewater using aluminium enriched ferric oxide-hydroxide recovered from authentic acid mine drainage J. Hazard. Mater., 2017, 325, 223-229.

[16] Symonds, M.R.E., Moussalli, A. A brief guide to model selection, multimodel inference and model averaging in behavioural ecology using Akaike's information criterion. Behav Ecol Sociobiol 65, 13-21

[17] MontoyaV.H., ServinJ.G., "Thermal Treatments and Activation Procedures used in the Preparation of Activated Carbons" Instituto Tecnológico de Aguascalientes México, 2(92), 2012.

[18] S. I. Siddiqui and S. A. Chaudhry, Iron oxide and its modified forms as an

adsorbent for arsenic removal: A comprehensive recent advancement Process Saf. Environ. Prot., 2017, 111, 592-626

[19] K. Jomova, Z. Jenisova, M. Feszterova, et al., Arsenic: toxicity, oxidative stress and human disease J. Appl. Toxicol., 2011, 31, 95-107.

[20] L. Fang, X. Y. Min, R. F. Kang, et al., A critical review on arsenic removal from water using iron-based adsorbents Sci. Total Environ., 2018, 639, 110-117.

[21] M. A. Hoque, W. G. Burgess and K. M. Ahmed, Itregation of aquifer geology, ground water flow and arsenic distribution in deltaic aquifers- a unifying concept Hydrol. Processes, 2017, 31, 2095-2109.

[22] S. V. Jadhav, E. Bringas, G. D. Yadav, et al., Arsenic and fluoride contaminated groundwaters: a review of current technologies for contaminants removal J. Environ. Manage., 2015, 162, 306-325.

[23] M. Singh, A. K. Singh, N. Srivastava, et al., Arsenic mobility in fluvial environment of the Ganga Plain, northern India Environ. Earth Sci., 2010, 59, 1703-1715

[24] W. Baeyens, A. de Brauwere, N. Brion, et al., Arsenic speciation in the River Zenne, Belgium Sci. Total Environ., 2007, 384, 409-419.

[25] A. Barats, G. F'eraud, C. Potot, et al. Naturally dissolved arsenic concentrations in the Alpine/ Mediterranean Var River watershed (France), Sci. Total Environ., 2014, 473, 422-436.

[26] M. Arain, T. Kazi, J. Baig, et al., Co-exposure of arsenic and cadmium through drinking water and tobacco smoking: Risk assessment on kidney dysfunctionFood Chem. Toxicol., 2009, 47, 242-248.

[27] Andjelkovic, S. Azari, M. Erkelens, et al., Bacterial iron-oxide nanowires from biofilm waste as a new adsorbent for the removal of arsenic from water† RSC Adv., 2017, 7, 3941-3948.

[28] V. Pallier, G. Feuillade-Cathalifaud, B. Serpaud, et al., Effect of organic matter on arsenic removal during coagulation/flocculation treatment J. Colloid Interface Sci., 2010, 342, 26-32.

[29] G. Ungureanu, S. Santos, R. Boaventura, et al., Effects of iron oxide on antimony(V) adsorption in natural soils: transmission electron microscopy and X-ray photoelectron spectroscopy measurements J. Environ. Manage., 2015, 151, 326-342.

[30] S. Maeda, S. Nakashima, T. Takeshita, and S. Higashi, "Bioaccumulation of arsenic by freshwater algae and the application to the removal of inorganic arsenic from an aqueous phase. Part II. By *Chlorella vulgaris* isolated from arsenic-polluted

[31] J. Yoon, G. Amy, J. Chung, et al., Occurrence of Perchlorate in Drinking Water and Seawater in South Korea Chemosphere, 2009, 77, 228-235.

[32] W. Driehaus and F. Dupont, A critical review on arsenic removal from water using iron-based adsorbents Eur. J. Water Qual., 2010, 36, 138-147.

[33] M. A. Pintor, B. R. C. Vieira, S. C. R. Santos, et al., Drinking Water Treatments for Arsenic and Manganese Removal and Health Risk Assessment in White Rock, Canada Sci. Total Environ., 2018, 642, 1075-1089.

[34] P. Qi and T. Pichler, Chemosphere, 2016, Volume 145, 55-60.

[35] J. Cantu, L. E. Gonzalez, J. Goodship, et al., Removal of Arsenic from water using synthetic Fe7S8 nanoparticles Chem. Eng. J., 2016, 290, 428-437.

[36] M. Podder and C. Majumder, Optimization of environmental conditions for the growth and bioaccumulation of As(III) and As(V) ions by growing Corynebacterium glutamicum MTCC 2745: inhibition kinetic study Spectrochim. Acta, Part A, 2016, 153, 298-314.

[37] M. C. Collivignarelli, M. Canato, S. Sorlini, et al., What Advanced Treatments Can Be Used to Minimize the Production of Sewage Sludge in WWTPs?Water Practice & Technology, 2016, 11, 115-126.

[38] M. K. Ghosh, G. E. J. Poinern, T. B. Issa, et al., Arsenic adsorption on goethite nanoparticles produced through hydrazine sulfate assisted synthesis method Korean J. Chem. Eng., 2012, 29, 95-102.

[39] S. Vasudevan and J. Lakshmi, Electrochemical removal of boron from water: Adsorption and thermodynamic studies Can. J. Chem. Eng., 2012, 90, 1017-1026.

[40] R. Chowdhury, Using adsorption and sulphide precipitation as the principal removal mechanisms of arsenic from a constructed wetland – a critical review Chem. Ecol., 2017, 33, 560-571.

[41] NANZHU LI , Ferrate as a New T ate as a New Treatment Chemical for Remo eatment Chemical for Removal of Effluent al of Effluent Organic Matter (E ganic Matter (EfOM) and Emer fOM) and Emerging Micr ging Micro-Pollutants in T ollutants in Treated Municipal Wastewater for Water Reuse, Montclair State University Upper Montclair, NJ 2017

[42] M. C. Faria, R. S. Rosemberg, C. A. Bomfeti, et al., Removal of arsenic from water through adsorption onto metal oxide-coated material, Book, Applications of Adsorption and Ion Exchange Chromatography in Waste Water TreatmentVol. 15, pp 227-276, 2017.

[43] X. Sun, C. Hu, X. Hu, et al., Characterization and adsorption performance of Zr-doped akaganéite for efficient arsenic removal J. Chem. Technol. Biotechnol., 2013, 88, 629-635.

[44] W. Tang, Q. Li, S. Gao, et al., Arsenic (III, V) removal from aqueous solution by ultrafine α-Fe2O3 nanoparticles synthesized from solvent thermal method. J. Hazard. Mater., 2011, 192, 131-138.

[45] H. Park, N. V. Myung, H. Jung, et al., Electrochemical synthesis of magnetic iron oxide nanoparticles with controlled size J. Nanopart. Res., 2009, 11, 1981-1989.

[46] Mohammad-Khah A, Ansari R Activated charcoal: preparation, characterization and applications: a review article. Int J Chem Tech Resour, 2009, 1:859-864

[47] Ansari R, Sadegh M Application of activated carbon for removal of arsenic ions from aqueous solutions. E- J Chemistry, 2007, 4:103-108

[48] Bansal RC, Donet JB, Stoeckli F, A review of active carbon, 1988, Marcel Dekker, New York

[49] Kairvelu K, Thamaraiselvi K, Namazivayam C Removal of heavy metals from industrial wastewaters by adsorption onto activated carbon prepared from an agricultural solid waste. Bioresour Technol, 2001, 76:63-65

[50] Srivastava S, Agrawal SB, Mondal MK, A review on progress of heavy metal removal using adsorbents of microbial and plant origin, Environ Sci Pollut, 2015, 22:15386-15415

[51] Kairvelu K, Thamaraiselvi K, Namazivayam C, Removal of heavy metals from industrial wastewaters by

adsorption onto activated carbon prepared from an agricultural solid waste, Bioresour Technol, 2001, 76:63-65

[52] Lorenzen L, van Deventer JSJ, Land WM Factors affecting the mechanism of the adsorption of arsenic species on activated carbon, Miner Eng, 1995, 8:557-569

[53] Li Z, Deng S, Yu G, Huang J, Lim VC As(V) and As(III) removal from water by a Ce–Ti oxide adsorbent: behavior and mechanism, J Chem Eng,2010, 161:106-113

[54] Liu Z, Zhang FS, Sasai R, Arsenate removal from water using Fe3O4-loaded activated carbon prepared from waste biomass. J Chem Eng, 2010, 160:57-62

[55] Pattanayak J, Mondal K, Mathew S, Lalvani SB, A parametric evaluation of the removal of As(V) and As(III) by carbon based adsorbents, Carbon, 2000, 38:589-596

[56] Genc-Fuhrman TJC, McConchie D, Increasing the arsenate adsorption capacity of neutralized red mud (Bauxsol). J Colloid Interface Sci, 2004, 271:313-320

[57] Shafique U, Ijaz A, Salman M, UzzamaW, Jamil N, Rehman R, Javaid A, Removal of arsenic from water using pine leaves. J Taiwan Inst Chem Engineers, 2012 43:256-263

[58] Chuang CL, Fan M, Xu M, Brown RC, Sung S, Saha B, Huang BL, Adsorption of arsenic(V) by activated carbon prepared from oat hulls. Chemosphere, 2005 Chapter-21, Page no.-478-483

[59] Chang QG, Lin W, Ying WC, Preparation of iron-impregnated granular activated carbon for arsenic removal from drinking water,J Hazard Mater, 2010,184:515-522

[60] Tuna AOA, Ozdemir E, Simsek EB, Beker U, Removal of As(V) from aqueous solution by activated carbon-based hybrid adsorbents: impact of experimental conditions, J Chem Eng, 2013, 223:116-128

[61] Lodeiro P, Kwan SM, Perez JT, Gonzalez LF, Gerente C, Andres Y, McKay G, Novel Fe loaded activated carbons with tailored properties for As(V) removal: adsorption study correlated with carbon surface chemistry, J Chem Eng, 2013, 215:105-112

[62] Samsuri AW, Sadegh-Zadeh F, Seh-Bardan BJ, Adsorption of As(III) and As(V) by Fe coated biochars and biochars produced from empty fruit bunch and rice husk, J. Environ. Chem. Eng., 2013, 1: 981-988

[63] Chammui Y, Sooksamiti P, Naksata W, Thiansem S, Arqueropanyo OA, Removal of arsenic from aqueous solution by adsorption on Leonardite, JChem Eng, 2014,240: 202-210

[64] Lunge S, Singh S, Sinha A, Magnetic iron oxide (Fe3O4) nanoparticles from tea waste for arsenic removal, Magn Mater, 2014, 356:21-31

[65] Sigrist ME, Brusa L, Beldomenico HR, Dosso L, Tsendra OM, Gonzalez MB, Pieck CL, Vera CR, Influence of the iron content on the arsenic adsorption capacity of Fe/GAC adsorbents, J Environ Chem Eng, 2014, 2:927-934

[66] Chen RZ, Zhang ZY, Yang N, Lei ZF, ChenN, Guo X, Zhao C, Sugiura N, Use of ferric-impregnated volcanic ash for arsenate(V) adsorption from contaminated water with various mineralization degrees, J Colloid Interface Sci, 2011, 353:542-548

[67] Gerente C, Andres Y, McKay G, Le Cloirec P,Removal of arsenic(V) onto

chitosan: from sorption mechanism explanation to dynamic water treatment process, J Chem Eng, 2010, 158:593-598

[68] S. Mahimairaja, N. S. Bolan, D. C. Adriano, and B. Robinson, Arsenic contamination and its risk management in complex environmental settings, Advances in Agronomy, 2005, vol. 86, pp. 1– 82

[69] C. Sullivan, M. Tyrer, C. R. Cheeseman, and N. J. D. Graham, Disposal of water treatment wastes containing arsenic, Science of the Total Environment, 2010, 408, 1770-1778

[70] O. Atteia, E. D. C. Estrada, and H. Bertin, Soil flushing: the origin of efficiency variability, Review of Environment, Science and Biotechnology, 2013, 12, 379-389.

[71] A. A. L. S. Duarte, S. J. A. Cardoso, and A. J. Alcada, Emerging and innovative techniques for arsenic removal applied to a small water supply system, 2009, Sustainability, 1, 1288-1304.

[72] M. Koma'rek, A. Vane˘k, and V. Ettler, Chemical stabilization of metals and arsenic in contaminated soils using oxides, Envi- ronmental Pollution, 2013, 172, 9-22.

[73] N. C. Mueller and B. Nowack, Nanoparticles for remediation: solving big problems with little particles, Elements, 6, 395-400.

[74] K.-R. Kim, B.-T. Lee, and K.-W. Kim, Arsenic stabilization in mine tailings using nano-sized magnetite and zero valent iron with the enhancement of mobility by surface coating, Journal of Geochemical Exploration, 2012, 113, 124-129.

[75] E. D. Leiva, C. D. P. Ra'mila, I. T. Vargas et al., Natural attenuation process via microbial oxidation of arsenic in a high Andean watershed, Science of the Total Environment, 2014, 466- 467, 490-502.

[76] E. K. Porter and P. J. Peterson, Arsenic accumulation by plants on mine waste (United Kingdom), Science of the Total Environment, 1975, 4, 365-371.

[77] M. I. Silva Gonzaga, J. A. Gonzaga Santos, and L. Q. Ma, Arsenic phytoextraction and hyperaccumulation by fern species, Scientia Agricola, 63, 1, 2006, 90-101, 2006.

[78] Q. Yang, S. Tu, G. Wang, X. Liao, and X. Yan, Effectiveness of applying arsenate reducing bacteria to enhance arsenic removal from polluted soils by Pteris vittata L, International Journal of Phytoremediation, 2012, 14,89-99.

[79] J. Xiong, W. Wang, H. Fan, L. Cai, and G. Wang, Arsenic resistant bacteria in mining wastes from Shangrao coal mine of China, Environmental Science and Technology, 2006, 1, 535– 540.

[80] H. M. Sehlin and E. B. Lindstrom, Oxidation and reduction of arsenic by Sulfolobus acidocaldarius strain BC, FEMS Microbiology Letters, 1992, 93, 87-92.

[81] D. Lie'vremont, P. N. Bertin, and M.-C. Lett, Arsenic in contaminated waters: biogeochemical cycle, microbial metabolism and biotreatment processes, Biochimie, 2009, 91, 1229-1237.

[82] D. Muller, D. Lie'vremont, D. D. Simeonova, J.-C. Hubert, and M.-C. Lett, Arsenite oxidase aox genes from a metal-resistant β-proteobacterium, Journal of Bacteriology, 2003, 185, 135-141.

[83] K. Chitpirom, S. Tanasupawat, A. Akaracharanya et al., Coma- monas terrae sp. nov., an arsenite-oxidizing bacterium isolated from agricultural soil in Thailand, Journal of General and Applied Microbiology, 2012, 58, 245-251.

[84] Margarita Stoytcheva and Roumen Zlatev (July 25th 2018). Introductory Chapter: Arsenic, Arsenic - Analytical and Toxicological Studies, Margarita Stoytcheva and Roumen Zlatev, IntechOpen.

Chapter 4

Remedial Approaches against Arsenic Pollution

Gia Khatisashvili, Tamar Varazi, Maritsa Kurashvili, Marina Pruidze, Evgeni Bunin, Kakha Didebulidze, Tinatin Butkhuzi, Elina Bakradze, Nino Asatiani, Tamar Kartvelishvili and Nelly Sapojnikova

Abstract

The study is devoted to a very urgent and acute problem for Georgia – remediation/restoration of the arsenic (As) mining and storage sites. The approach of a given work is based on using capabilities of nature itself, which has a great adaptive potential to chemical environmental pollution. The aim of the study is to identify the bacterial strains from the endemic soil microbiota, characteristic to a specific localization of arsenic contaminated sites and able to resist to the toxicant. To determine the level of arsenic contamination, soil samples have been analyzed using Inductively Coupled Plasma - Optical Emission Spectrometry method. The distribution of arsenic in soil samples splits them into categories according to the degree of contamination, ranging from 50 ppm to 13000 ppm. The local bacteria community has been studied using conventional cultivation method along with modern method of bioindication – a biochip. The low density biochip contains the relevant probes for the identification of the bacterial consortium in soil microbiota. Chemical and microbiological analysis was based on the standards and methodologies developed by International Standards Organizations – ISO and Environmental Protection Agency – EPA. It is prospected that bioremediation can become essential part of remediation against arsenic pollution in the context of circular economy.

Keywords: arsenic pollution, inductively coupled plasma - optical emission spectrometry method, biochip, soil microbiota, bioremediation, PCR amplification

1. Introduction

The study deals with chemical pollution of the environment, in particular arsenic contamination of soil, which is a highly pressing problem for both Georgia and the world. Arsenic is one of the toxic metalloids that exist in more than 200 mineral forms, where 60% of them are normally arsenates, 20% are sulfosalts and sulfides and 20% are arsenite, oxides, arsenide, silicates and elemental arsenic. Arsenic mainly exists in the environment as arsine (As^{-3}), elemental arsenic (As^{0}), arsenite (As^{+3}) and arsenate (As^{+5}). Among all these forms only arsenite and arsenate are more abundant in natural environment than the other two [1]. Arsenic is naturally present in the lithosphere (earth crusts, soil, rock and sediment), hydrosphere (surface water, aquifers, deep well and oceans) and biosphere (food chain and ecosystems) [2, 3]. Arsenic is one of the top five toxic chemicals that were listed in the US Comprehensive Environmental Response, Compensation, and Liability (CERCLA) act of hazardous substances [4]. The toxicity of arsenic to living organisms is due to its functional affinity for phosphorus and its ability to form covalent bonds with sulfur. Arsenate replaces phosphate in phosphorylation reactions, and arsenite interacts with thiol groups of proteins, causing serious disturbances in cell metabolism [5]. Arsenic species are deposited in the skin, lungs, kidney, liver, etc. and cause several severe diseases by oxidative stress, altered DNA methylation, altered DNA repair, mitochondrial damage, cell proliferation, tumor promotion and co-carcinogenesis [6]. Arsenic exists both in toxic inorganic (associated with iron, cobalt, nickel coupled with sulfate minerals) and comprehensively less toxic organic (associated with carbon and hydrogen) forms. It is released into the environment by natural phenomenon (geogenic) or by anthropogenic activities. Soil texture is an important characteristic that affects arsenic chemistry. Arsenic levels in soil depend on climate, pH and redox potential. Arsenic can be chemically transformed in soils through several mechanisms. These include oxidation, reduction, adsorption, dissolution, precipitation, and volatilization. The trivalent form (As^{+3}) is easily absorbed to ferric and aluminum oxides in the soil environment and oxidized to the pentavalent form in aerobic condition and reduced back to the trivalent form by reduction [7]. Under aerobic environment, the inorganic form of arsenic can easily bind to inorganic and organic materials in soil such as clay, iron and manganese dioxide, and exists in the pentavalent state (arsenate AsO_4^{3-}). Under anaerobic environment, anaerobic bacteria transform it into less toxic volatile forms such as dimethyl arsenic acid and monomethyl arsenic acid [8]. Oxidation and reduction of arsenic species are carried out both chemically and biologically in soil and water. The acid/base chemistry plays a major role in the types of arsenical compounds present in the soils. Naturally, arsenic can be released into the soil environment by weathering and erosion (hydrolysis and oxidation process) of primary sulfide mineral (arsenopyrite). Organic content, iron and aluminum oxides, hydroxides, and phosphate ion play an important role in the speciation as well as mobility and sorption of arsenic in sediments [9]. Arsenic retention in soil media is mainly dictated by adsorption and desorption reactions, presence of ligands and soil redox conditions [10]. Bioavailability plays an important role in uptake of arsenic from soil to plants. The widespread contamination of the inorganic form of arsenic and its level of toxicity is becoming a global problem as the metalloid does not degrade, and cannot be destroyed. It is reported that conventional methods such as oxidation or reduction, chemical precipitation, filtration, ion exchange, reverse osmosis and evaporation recovery of cleaning contaminated area are too expensive and laborious. There is need to develop eco-friendly and low cost technique to mitigate the arsenic contamination. It is well documented that bioremediation is a cost-effective and a comparatively innocuous alternative to physical methods for heavy-metal remediation [11]. Wide varieties of microorganisms are capable of growth in presence of heavy metal ions and can tolerate high concentrations [12]. Although arsenic is generally toxic for life, microorganisms can use arsenic compounds as electron donors, electron acceptors or show arsenic detoxification mechanisms [13]. Furthermore, suggestions have been made about the existence of microorganisms in which arsenic can play the role of phosphorus. These microorganisms are able to live and reproduce in conditions of phosphorus deficiency, replacing phosphorus in DNA with arsenic toxic to other life forms [14]. Various bacteria such as *Acidithiobacillus*, *Bacillus*, *Deinococcus*, *Desulfitobacterium* and *Pseudomonas* have been reported to be resistant to arsenic [15–17]. It has been reported that the strains of *Aeromonas*, *Exiguobacterium*, *Acinetobacter*, *Bacillus* and *Pseudomonas* can tolerate high concentrations of arsenic up to 100 mM arsenate or up to 20 mM arsenite [18]. Molecular-biological studies have shown that they possess arsenic resistance genes. *ArsA* and *arsB* genes encode transmembrane pumps that remove As^{+3} from the cytoplasm, reducing the concentration of arsenic in the cell. The *arsC* gene encodes an enzyme capable of transforming As^{+5} into As^{+3}. The *arr* gene encodes a periplasmic As^{+5} reductase, which uses As^{+5} as an electron acceptor during anaerobic respiration. The *aox* gene encodes a periplasmic As^{+3} oxidase that oxidizes As^{+3} to As^{+5}. The genes *arsR* и *arsD* are regulatory genes [19, 20]. The ecosystems near arsenic mining industrial areas are characterized with an elevated level of pollutants in Caucasus region; such hot spots are presented in Western Georgia (Uravi, Tsana) abandoned arsenic production facilities and nearby mining tailings are stored in deteriorating conditions that pose a threat to the population. In this study, we evaluated the diverse populations and functioning microbial communities in the soil for natural and external remediation in environmental cleanup; the level of As contamination in the wide area of Western Georgia hot spots; the characteristics of microorganisms cultivated from the highly contaminated regions.

2. Biochip applications

The establishment of the structure of the collective genomes (metagenomes) of environmental microbial communities (microbiomes) has a decisive influence on the development of earth sciences of the 21st century. However, it soon has become apparent that gene sequence data alone is of relatively little use unless it was directly linked to agriculture, alternative energy production, industrial processes, and environmental cleanup relevance. Monitoring of bacterial consortium in contaminated regions will enable to assess in advance the extent of native bioremediation and accordingly suggest a strategy of external bioremediation (detoxification). Diagnostic biochip is able to enumerate bacteria, involved in bioremediation. A biochip is a platform for screening a single sample for multiple purposes. The constructed and applied in the presented study low density DNA-biochip is a small solid matrix with multiple dot markers (probes). Each marker contains genetic DNA-sensors (pico-amounts of oligonucleotides), which is designed according to the genetic map of the studied objects.

2.1 Biochip design: DNA probes and biochip matrix

Effect of toxic metals in natural conditions is rarely revealed by the action of one of them, mainly it is a complex action of several metals (sometimes even non-toxic), and the same concerns *arsenic* contaminated area. The groups of bacteria that can be involved in the arsenic and heavy metals transformation were used as targets for a low-density biochip. The constructed low-density biochip contains 16 probes in duplicate (**Table 1**). Probes 1–3 are derived from 16S rRNA genes. All other probes (4–16) are derived from functional genes. Highly specific probes covering the taxa of interest were selected for subsequent testing from the publications pointed in **Table 1**. The BLAST NCBI database (http://www.ncbi.nlm.nih.gov) confirms bacterial specificity for a probe sequence. A crucial step in biochips design is the evaluation of probes hybridization capacity. Even though specific probes for target microorganisms are generated following well-established requirements, this would require an experimental confirmation of their equal hybridization capacity. The experimental cassette approach proves the selected probes equal hybridization capacity [22, 26]. Therefore, the variations in the fluorescent intensities on a biochip, coming from probe's size and nucleotide sequence differences are excluded. The equalized biochip is the basis for the estimation of the microbial proportion in the consortium, assuming that bacterial functional genes are presented in one

N	Probe name	Target species	Gene name	Sequence 5′ → 3′	Ref.
1	Archaea1	*Archaea*	16S rRNA	GTG CTC CCC CGC CAA TTC AT	[21]
2	GeoS	*G. sulfurreducens*	16S rRNA	TTC GGG CCT CCT GTC TTT C	[22]
3	GeoM	*G. metallireducens*	16S rRNA	TTC GGG CCT TTT GTC ACC	[22]
4	BacilPhosL	*Bacillus* spp.	Phospholipase PL	CTA CTG CCG CTC CAT GAA TCC	[23]
5	Pseudomonas	*Pseudomonas* spp.	Polyketide synthase phlD	GAG GAC GTC GAA GAC CAC CA	[24]
6	napA	*Nitrate Reducing Bacteria*	nitrate reductase (napA)	CCG CGG CTA TGT GGG TCG AAA AAG	[22]
7	nirS	*Nitrate Reducing Bacteria*	nitrite reductase (nirS)	CGC TGT TCG TCA AGA CCC ATC CG	[22]
8	Clostridia	*Clostridia*	Fe-hydrogenase hydA	CGG CGA GCA TGA TCC AGC AAT	[22]
9	Shew	*Shewanella*	Ni,Fe-hydrogenase hydA	ACA ACT GCC CAA CCG AGC G	[22]
10	FTHFS	*Acetogenic bacteria*	formyltetrahydrofolate synthetase (fthfs)	TGC ATG GCC AAG ACC CAA TAC AGC	[22]
11	a/bssA	*Hydrocarbon-degrading bacteria*	alkylsuccinate synthase and benzylsuccinate synthase alpha subunits (assA/bssA)	TCG TCA TTG CCC CAT TTG GGG GC	[22]
12	SRB	*Sulfate-reducing bacteria*	adenosine 5′-phosphosulfate-reductase APR	CCA GGG CCT GTC CGC CAT CAA TAC	[25]
13	Dsmicrobium1	*Desulfomicrobium*	hydA (Ni,Fe-hydrogenase)	CCA CAA CCT GGC CAT CCC GGA AAT	[22]
14	Dtomaculum2	*Desulfotomaculum*	hydA (Fe-hydrogenase)	CAC GCA TCG GGG AGA GGG TGG	[22]
15	Dbulbus	*Desulfobulbus*	hydA (Ni,Fe-hydrogenase)	GCG CCA CCC TGC CGT TCA AC	[22]
16	Dbacterium1	*Desolfobacterium*	hynB (periplasmic Ni,Fe hydrogenase)	CAC TGG AAC AGG CGA TCA AG	[22]

Table 1.
Characteristics of probes used in this study.

copy. As the 16S rRNA genes are multiplied in the bacterial genome, the probes 1–3 are not included in the consideration of the bacterial proportion in the samples. The signals from the probes 1–3 can only confirm or disprove the presence of the target species in the soil samples and can show the comparative relation between the studied samples. Biochip dendrimeric matrix technology and conditions of hybridization on a biochip were the same as described previously [26].

2.2 Analysis of bacterial composition using a biochip

The intensive study of heavy metal contaminated areas has revealed special property of bacteria living in these areas. It is their selective metallophilic property. That is largely due to the activity of these organisms to change the oxidation state of soluble heavy metal oxyanions by transformation them to an insoluble form. Metal-reducing bacteria may receive metal ions as soluble oxyanions via the anionic transport system. Some metal-reducing bacteria can play a key role in the mobilization of arsenic in sediments collected from a contaminated aquifer [27]. *Shewanella*, *Clostridia*, *Pseudomonas* spp., *Bacillus* spp., *Geobacter* spp., *Archaea* exhibit metal-resistant, metal-reducing properties. Sulfate reducing bacteria (SRB) are metal-tolerant. They show the potential to remove As and other metals from As mining area [28]. Active nitrate-reducing population in soil may oppose to SRB activity to produce hydrogen sulfide, the key metal precipitating substrate. Activity of acetogenic bacterial population is a source of energy and carbon.

To evaluate and compare bacterial composition of arsenic contaminated area soil profiles were sampled at different points. Three sampling points were selected for a biochip analysis. Samples from the arsenic ore collector area were selecting based on the location: soil from arsenic processing plant (**P**) sample (GPS 42.66760°N, 43.30048°E) and refers to artificial primitive soil; soil from arsenic ore (**O**) sample (GPS 42.66768°N, 43.30021°E) and refers to artificial primitive soil. The sample (GPS 42.62343°N, 43.34136°E) is considered as control sample (**C**), located 6 km far from the ore collector area and refers to brown forest soils. All these soil samples (20–25 cm depth) were collected using sterile materials in hermetic plastic 50 ml flasks, transported to the laboratory at 4°C, and stored at - 20°C. Conditions of soil DNA preparation were the same as described previously. The steps of DNA fragmentation and its fluorescent labelling are very important in the process of biochip visualization and precede the hybridization of DNA fragments with the immobilized probes onto the matrix for the signal registration [22]. The intensity of the fluorescent signal on a biochip is proportional to the level of the bacteria in the assemblage.

2.2.1 Comparison of bacterial functional assemblages

The **Figure 1A** represents the hybridization signal intensities estimated as signal to noise (S/N) ratio for the probes targeting key genes encoding enzymes involved in metabolic processes. Functional gene array is the approach to assess diversity and functional activities of microbial communities in natural environments. The **Figure 1A** expresses the proportion of the issued bacterial species in the samples **C, P** and **O** functional assemblage. The DNA of the control (**C**) sample showed the highest level of the fluorescent signals with the functional gene probe targeting *Bacillus* spp. The pointed bacterial species dominate in the studied bacterial consortium. The DNA of the samples from arsenic processing plant (**P**) and from arsenic ore (**O**) revealed the significant increase of the fluorescent signals with the probes targeting metal-tolerant species *Bacillus, Shewanella, Clostridia*. *Bacillus* spp. are using As^{+5} as an electron acceptor and as a result arsenate is reduced to arsenite. The populations with nitrate/nitrite reduction and acetogenic activities increased dramatically as well.

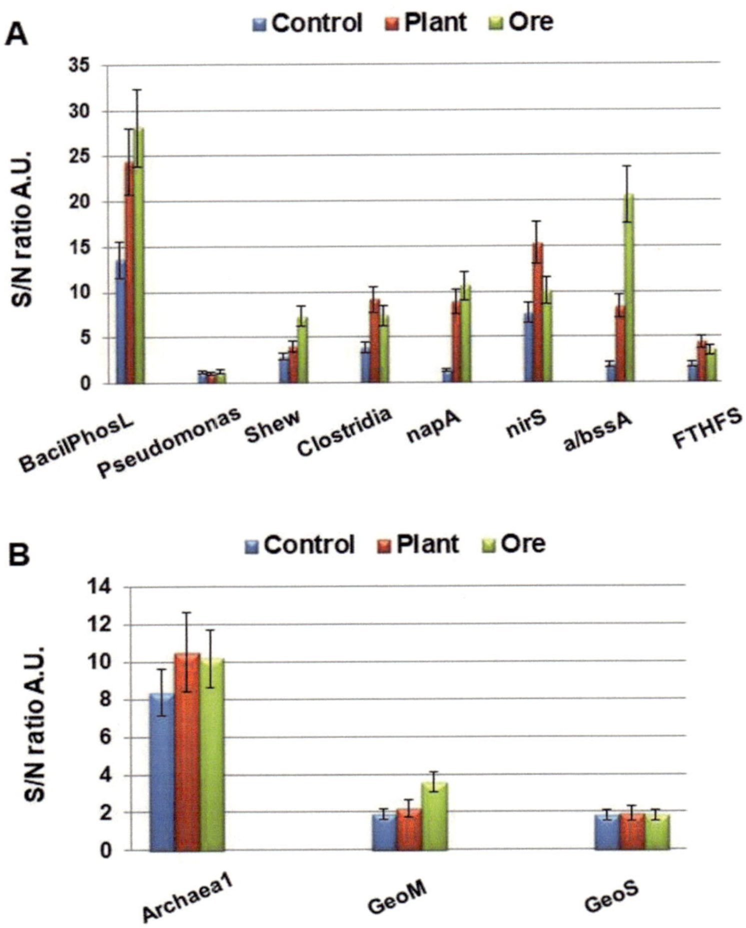

Figure 1.
Comparative analysis of the bacterial profiles of non-contaminated (Control) and arsenic contaminated (Plant) and (Ore) soils, performed by DNA hybridization on biochip. The results of the hybridization are presented as a signal to noise (S/N) ratio. The data presented are mean values ± SD from three separate sets of experiments

And only the *Pseudomonas* species remain at background levels at all studied areas. If *Pseudomonas* are present in small amounts in soil, it can be said that the content As^{+3} is low in this soil because these bacteria receive metabolic energy by its oxidation.

The hybridization signals from the probes targeting 16S rRNA genes are presented in **Figure 1B**. The presence *Archaea* and *Geobacter* spp. in all samples is confirmed by the fluorescent signals from the subsequent probes (Archaea1, GeoM, and GeoS). However, their content does not differ between the control and arsenic processing areas.

2.2.2 Comparison of sulfate reducing bacteria functional assemblages

Sulfate reducing bacteria take the special attention, as SRB along with methanogens are involved in arsenic methylation and demethylation in soils [29], and microbial arsenic methylation and demethylation are important components of the As biogeochemical cycle. Only a limited number of organisms are well known for their narrowly defined metabolic capabilities, and SRB is such group. The detection of SRB in the bacterial consortium could be carried through functional genes as *dsr* and *apr* (genes directly associated with the reduction of inorganic sulfate). We use the probe (SRB) based on the highly homologous sequence of *apr* gene as the universal probe for the detection of the wide spectra of SRB [25]. In addition, we expanded the detection of different SRB species using the probes based on the hydrogenase genes for the specific recognition of *Desulfomicrobium*, *Desulfobacterim*, *Desolfotomaculum* and *Desolfobulbus* species in the SRB assemblage. The results of SRB detection in the selected samples (**C, P,** and **O**) are presented in **Figure 2**. **Figure 2A** presents the biochip results. **Figure 2B–E** present the result of PCR amplification of the soil DNA with the primers specific for *Desulfomicrobium*, *Desulfobacterim*, *Desolfotomaculum* and *Desolfobulbus* species on the 1.5% agarose gel (1xTAE running buffer, ethidium bromide staining). M - Gene Ruler 100 bp Plus DNA Ladder (Thermo Scientific, USA), C+ – positive control.

PCR primers are listed in **Table 2**. Individual gene-specific PCR primers were designed using Primer3Plus software. The PCR mixture (20 μl) contained 1xiProof High-Fidelity Master Mix (Bio-Rad Laboratories, USA) with 1.5 mM $MgCl_2$, 200 μM (each) deoxynucleoside thriphosphate, 500 nM each primer, 20 ng of DNA template. PCR amplification was conducted using the following conditions identical for each primer pair: an initial denaturation step (30 s, 98°C) was followed by 30 cycles of denaturation (10 s, 98°C), annealing (20 s, 60°C), and extension (15 s, 72°C) and one terminal extension step (10 min, 72°C).

As it follows from the analysis of the SRB assemblages, the use of nothing but the wide spectra probe (SRB) on a biochip does not differentiate the content of SRB in the studied samples. The expansion of the SRB allows distinguishing SRB composition between the samples. Specifically, *Desolfotomaculum* and *Desolfobulbus* spp. are partly inactivated in arsenic processing areas (**P** and **O**) in comparison with the control

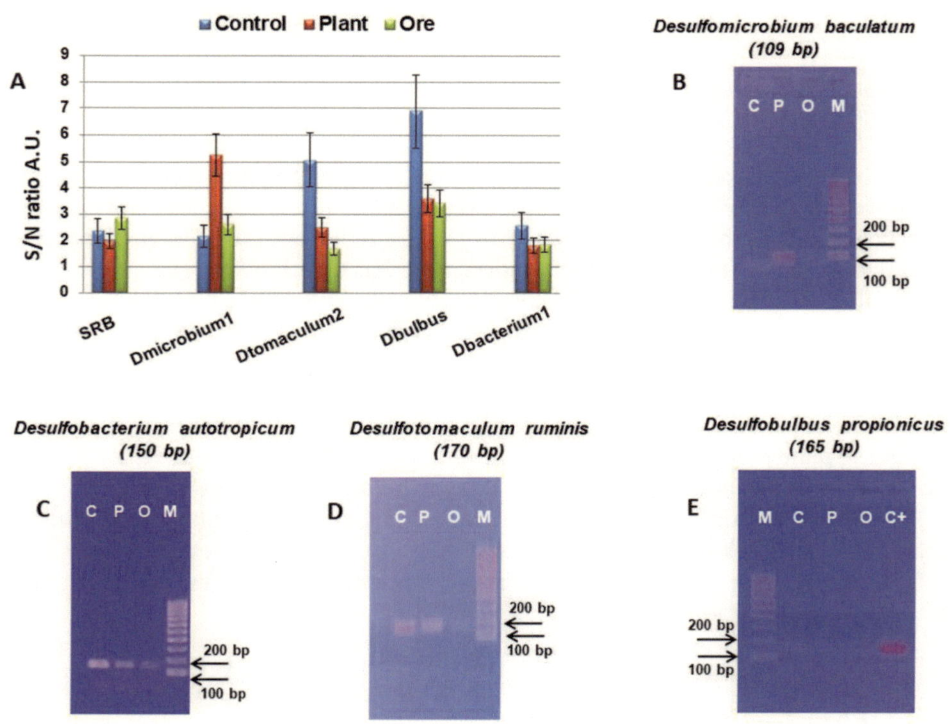

Figure 2.
Comparative analysis of sulfate reducing bacteria profiles, performed by DNA hybridization on biochip (Panel A) and by PCR amplification (Panels B, C, D, E).

Positive control DNA	Target gene GenBank accession no.	PCR Primers name	*Sequence* *5′→3′*	*PCR fragment size (bp)*
Desulfobacterium autotropicum DSM 3382	*hynB* (periplasmic Ni,Fe hydrogenase) CP001087	For2	*CTT GAG ACC ATT TCG GTT GA*	150
		Rev2	*GGC GCC GTT ATA GGC TGT AG*	
Desulfobulbus propionicus DSM2032	*hydA* (Ni,Fe-hydrogenase); CP002364	For1	*CTC TAC AAA CTG GGG TGC AAG*	165
		Rev1	*GAC CTT TGC CTT GGA AAA CA*	
Desulfomicrobium baculatum DSM 4028	*hydA* (Ni,Fe-hydrogenase); CP001629	For1	*GTT TCG CCG AAG AAC ATG A*	109
		Rev1	*CGC CCA ATC CCT ACA ACC T*	
Desulfotomaculum ruminis DSM 2154	*hydA* (Ni,Fe-hydrogenase); CP002780	For2	*GGA AAT TCC GGA CTG GTA CA*	170
		Rev2	*GAG GGA GTC TTC TCC AAG CA*	

Table 2.
DNA and PCR primers.

sample (**C**). Inhibition of some SRB species correlates with the dramatically increased nitrate-reducing activity in the corresponding samples, as nitrite (product of nitrate reductase) inhibits the enzyme dissimilarory sulfite reductase [30]. *Desolfobacterium* spp. are at the same level in all three samples. As it was mentioned above, *Desolfobacterium* spp. are among the species, which resist As. Only *Desulfomicrobium* spp. are significantly increased in the soil from arsenic processing plant (**P**). The biochip data coincides with the PCR data. The following elemental analysis, determining the total As content in the selected samples: **C** – 140 mg/kg soil; **P**– 6300 mg/kg soil; **O** – 110 mg/kg soil, points to the highest concentration of the As in the arsenic processing plant soil (**P**). The increased *Desulfomicrobium* spp. content can be considered as an indicator of highly polluted place even before the elemental analysis.

Taking into consideration biochip data about bacterial composition assessment, it can be said that a biochip is not only diagnostic, but also a predictive instrument.

The detailed analysis of total As content in the contaminated area is presented in the next Section.

3. Total As content analysis

Arsenic contamination in Georgia has many natural and anthropogenic sources. Arsenic ore extraction, processing and production of arsenic-containing preparations have been carried out on the territory of Georgia for decades. The processing of the ore, which is located in Ambrolauri region, started in 1937. Main products of processing were metallic arsenic of high purity, As_2O_5, As_2S_5 and tin arsenate. Technological cycle in factory was very simple and included thermal treatment of ore. Amount of waste is about 60 tons, which contains up to 1% As_2O_5. Waste was not used and it was stored at Kajiani territory in special hydrotechnical building, called tailings. The mining of ore was carried out at Lukhuni ore, processing was performed in the factory nearby village Uravi. Chemical mining factory in village Tsana started working is 1938. The main products were metallic arsenic and refined "white arsenic" (As_2O_3) (I grade – 99.9%, II grade – 99.5%). This substance is poorly soluble in water and permanently can have great negative influence on environment. The factory in Tsana, administrative buildings and warehouse farming are fully destroyed and collapsed. There is no fence around the territory. Arsenic kilns and containers are taken from the ground. Thus the problem is arsenic containing waste and soils. Until today the great amount of toxic waste of arsenic production is stored in villages Uravi and Tsana, near the territory of the factories (more than 120 000 tons waste, containing 4–9% of white arsenic), which is not located safely and there is a high risk of ecological disaster in rivers and soils, especially risks of natural disasters (floods, rockslide, erosion and etc.). It should be mentioned that LEPL National Environmental Agency is conducting permanent monitoring in order to determine arsenic content. The data are not encouraging [31]. Based on this it is necessary to determine arsenic content in ecosystems (soil, water) of this region. The fifty four soil samples encompassing two hectares have been analyzed using Inductively Coupled Plasma - Optical Emission Spectrometry (ICP-OES) method. Preparation (drying, loosening, sieving, extraction, etc.) of soil samples for pre-treatment and chemical analysis was carried out - ISO-11464; EPA 3050; According to EPA-TCLP-1111 ST methodologies. The results are presented in **Figure 3**. According to **Figure 3**, As content ranges from 47 mg/kg soil to 13000 mg/kg soil.

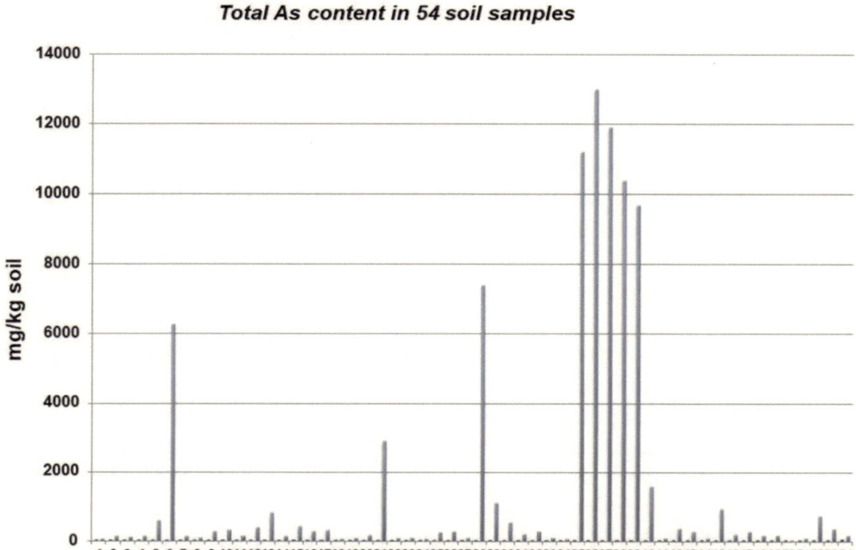

Figure 3.
Total arsenic content in 54 soil samples.

4. Characteristics of bacterial isolates from As highly contaminated area

4.1 Conditions of soil samples treatment for bacterial isolates recovery

Soil samples for bacterial isolates recovery were taken from the arsenic contaminated area and were selecting based on the location: soil sample from arsenic storage area (**SA**) (GPS 42.81425°N, 43.11586°E) and refers to brown forest black soil; soil sample from the place nearby the storage containers (**SC**) (GPS 42.81403°N, 43.11567°E) and refers to brown forest black. The sample (GPS 42.62343°N, 43.34136°E) is considered as control sample (**C**), located 6 km far from the storage area and refers to brown forest soils. The ICP-OES method determined the total As content in the selected samples: **C** – 140 mg/kg soil; **SA** – 7400 mg/kg soil; **SC** – 10400 mg/kg soil. Therefore, the bacterial isolates recovered from **SA** and **SC** samples are from highly contaminated area. Soil samples in the sterile tubes were first thermally treated at 80°C for 10 minutes. Then the samples were processed according to the appropriate methodology, which involves pre-incubation the soil bacteria, both gram-positive and gram-negative, in universal and selective growth medium. In particular, to stimulate the growth of gram-negative microorganisms, MacConcey Broth was used, in which 1 g of soil samples were incubated at different conditions, namely at 4°C, 20–25°C (room temperature), and 42°C for the particular stimulation of growth and further isolation of genus *Shewanella* microorganisms [32]. Three different types of growth medium were prepared for the isolation of bacterial strains. The composition of the first liquid medium was as follows (g/L): $Na_2HPO_4x7H_2O$ - 12.80; $KH_2PO_4x7H_2O$ - 3.00; NH_4Cl - 1.00, yeast extract - 2.00; CH_3COONa - 8.20; the second liquid medium was (g/L) LB broth - 10, yeast extract - 5; NaCl – 10; solid nutrient agar was used as the third growth medium. To determine arsenic respiration of the isolates, As^{+3} or As^{+5} salts were added to the growth medium [33]. Specific growth media M2 and M6 were prepared for this purpose. The composition of M2 medium was as follows (g/L): lactate – 0.45; NaCl – 1.17; KCl – 0.30; NH_4Cl – 0.15; $MgCl_2x6H_2O$ – 0.41, $CaCl_2$– 0.11, KH_2PO_4–0.20; Na_2SO_4–0.07, $NaHCO_3$–2.00, $NaAsO_2$ or Na_2HAsO_4–0.007. The composition of M6 medium was as follows (g/L): $MgSO_4$–1.00; NH_4Cl – 1.00; Na_2SO_4–1.00; K_2HPO_4–0.10; $CaCl_2$–0.05; lactate – 4.00; $FeSO_4$–0.002; $NaHCO_3$–8.00, arsenate or arsenite – 0.00. Incubation of 39 primary unidentified strains under anaerobic conditions at 31°C for 72 hours was performed in M2 and M6 growth media [34]. Microbial masses grown on solid and liquid medium were tested for arsenic oxidation–reduction ability by potassium permanganate method [35]. Nissui Compact Dry diagnostic-selective media were also used for primary screening of microorganisms.

4.2 Microorganisms properties

Bacterial isolates that were motile, had oxidative metabolisms, were oxidase and catalase positive, ornithine decarboxylase positive, and DNase positive, and produced H_2S on triple sugar iron slants within 72 h of incubation were identified as belonging to the phenospecies, respectively based on acid production from sucrose and maltose, growth on SS agar, and growth in the presence of 6.5% NaCl. All tests were performed according to the manufacturer's instructions. Routine biochemical test results were read daily for 72 h; oxidation of various carbohydrates was assessed by Liofilchem® Microbial Identification system after 7 days of incubation. Enzymatic plate assay results were read daily for 7 days; appropriate positive and negative control strains were included for each assay. Some additional enzymatic activities (2 h) and

DOI: http://dx.doi.org/10.5772/intechopen.98779

Soil sample	Determined microorganism	Strain	Catalase test	Oxidase test	Mannitol fermentation	Starch hydrolysis	Nitrate reduction test	Urease test	Indole test	Citrate utilization test	TSI agar test	Gram staining
C	*Bacillus* spp.	#01	p	p	p	p	n	n	n	p	alk/acid/alk	p
		#02	p	p	p	p	p	n	n	p	alk/acid	p
		#03	p	p	p	p	n	p	n	p	alk/alk/H_2S	p
		#04	p	p	p	n/d	n	p	n	n/d	alk/alk/H_2S	p
SA	*Bacillus* spp.	#05	p	p	p	p	n	n	n	n/d	alk/alk/H_2S	p
		#06	p	p	p	p	p	n	n	n	alk/acid	p
SC	*Bacillus* spp.	#07	p	p	p	n/d	n	n	n	n	alk/acid	p
		#08	p	p	n/d	p	p	p	n	p	alk/acid	p
		#09	p	p	p	p	p	n	n	p	alk/acid	p
		#10	p	p	p	p	p	n	n	n	alk/acid	p
		#11	p	p	p	p	n	n	n	n	alk/alk/H_2S	p

p - positive, n - negative. n/d - not determined, alk – alkaline.

Table 3.
Biochemical activity of Bacillus *isolates.*

Soil sample	Determined microorganism	Strain	Catalase test	Oxidase test	Mannitol fermentation	Starch hydrolysis	Nitrate reduction test	Urease test	Indole test	Citrate utilization test	TSI agar test	Gram staining
SA	*Shewanella* **spp.**	SH01	p	p	p	n/d	p	n	p	p	alk/ alk/ H_2S	n
		SH04	p	p	n/d	n/d	n	n	n	n	alk/ acid/	n
SC	**Non-specified strains**	X01	n	p	n/d	n/d	n	n	n	p	alk/ alk	n
		X02	n	p	n/d	n/d	n	n	n	p	akl/ alk	n
		X06	p	p	n/d	n/d	p	n	n	n	alk/ alk/ H_2S	p
		X07	p	p	n/d	n/d	p	n	n	n	alk/ alk/ H_2S	n

p - positive, n - negative. n/d - not determined, alk – alkaline.

Table 4.
Biochemical activity of Shewanella *and non-specified isolates.*

gelatinase activity (3 days) were estimated using Wee-Tab tablets or gelatin strips (Key Scientific Products, Roundrock, Tex.). Microorganisms have been studied for intracellular and extracellular enzymatic activities such as Oxidase test, Catalase test, Carbohydrate utilization test, Urease test, Indole test, Esculine test, Nitrate reduction test, Citrate utilization test, Gelatin hydrolysis test and Motility test.

4.3 Bacterial isolates identification and characterization

The screening of the bacterial composition using a biochip (Section 2) showed that sulfate-reducing bacteria, along with the representatives of *Bacillus, Clostridium* and *Shewanella* spp. predominate and are characterized by high content in the biomass of microorganisms from the As contaminated area. Laboratory cultivation in the above-mentioned media shows those gram-positive and gram-negative microorganisms, revealing sulfate-reducing activity, as *Bacillus cereus*, *Bacillus subtilis*, *Clostridium* spp., *Enterobacter* spp., and *Shewanella* spp., which are prevailing in the studied biomass.

The biochemical characteristics of selected *Bacillus* isolates are presented in **Table 3**. As follows from **Table 3**, the vast majority of *Bacillus* strains show catalase activity, all strains show pronounced oxidase activity. Tests of carbohydrates fermentation show that part of the *Bacillus* spp. strains can completely utilize carbohydrates. Inexplicable data were obtained when analyzing the results of citrate utilization and urease conversion tests. Genus *Bacillus* is known as citrate consumers and has no urease activity, but repeated analysis of the tests reveals that *Bacillus* isolates # 03, # 04 and # 8 are characterized by urease activity, which in our opinion is an exception. A similar trend was observed for the citrate utilization test data, namely, *Bacillus* isolates # 06, # 07, # 10 and # 11 did not utilize citrate and the diagnostic area did not change color. It should be mentioned that the full results of the tests for *Bacillus* isolates #04, #05 could not be evaluated; what may have been caused by the polymorphism of the microorganisms and the altered biochemical properties. The results of gelatin hydrolysis tests of microorganisms of the genus *Bacillus* showed that the vast majority of the selected microorganisms are characterized by the ability to hydrolyze gelatin.

The biochemical characteristics of *Shewanella* and non-specified isolates are presented in **Table 4**. *Shewanella* isolates # SH01, # SH04 and also isolates #X06, #X07 reveal positive both oxidase and catalase activities. In the next phase of the study, the selected microbial isolates were tested for arsenic oxidation–reduction ability. The results of arsenic respiration are presented in **Table 5**.

As follows from **Table 5**, the *Shewanella* isolates #SH01 and #SH04, as well as non-specified isolates #X01, #X02, #X06 and #X07, can oxidize As^{3+} to As^{5+} in the

Soil sample	Determined microorganism	Strain	Arsenic reduction		Arsenic oxidation	
			M2	M6	M2	M6
SA	*Shewanella* spp.	SH01	n	n	p	n
		SH04	n	n	p	n
SC	Non-specified strains	X01	n	n	p	n
		X02	n	n	p	p
		X06	n	n	p	n
		X07	n	n	p	n

Table 5.
Arsenic respiration by isolated strains.

studied growth conditions. Namely, the isolates exhibit this property in M2 growth medium. At this time the toxic form of arsenic transforms into a less toxic one. This result is very important for the planning and implementation of phytoremediation of arsenic-contaminated soil, because the presence of this microbiota in the soil will facilitate the growth of plants and their absorption of arsenic.

5. Conclusions

Evaluation of arsenic pollution and health risks, prevention of natural soils and waters from pollution in order to provide ecosystem enhancement and population safety is extremely important for the reality of Georgia. Arsenic as the most heavy metals are scattered and unevenly located in the earth's crust. Under natural conditions, in sufficiently high concentrations, it is found in areas of ore deposits. Arsenic toxicity is governed by its high affinity to sulfur and phosphorus, what is extremely dangerous for living organisms. In modern ecological biotechnologies aimed at cleaning up a chemically polluted environment, both from an economic and an ecological point of view, phytoremediation is considered the most effective, which consists in the purposeful planting of specially selected plants in contaminated areas. To use phytoremediation, it is necessary to study not only the level of pollution, but also the physicochemical properties of the soil and, mostly important, the soil microbiota. This is due to that a certain consortium of microorganisms has already formed in the contaminated soil, which, on the one hand, is accustomed to being present in a polluted environment, on the other hand, determines the conditions for plant growth on this soil and, in some cases, leads to the transformation of pollutants. Therefore, the screening of the indigenous bacteria was the first key step on the way of the bioremediation of arsenic contaminated area. Diagnostic biochip is used for the monitoring of the indigenous bacteria. The bacterial composition revealed the prevalence of metal-tolerant species as *Bacillus, Shewanella* and *Clostridia* in the arsenic contaminated area. The bacterial interrelation, specifically the significant increase of *Bacillus* spp., *Desulfomicrobium* spp. and a very low level of *Pseudomonas* spp. could indicate that As^{+5} is the predominant arsenic form in the studied sites, as *Bacillus* spp. use As^{+5} as an electron acceptor in bacterial respiration, and *Pseudomonas* spp. may receive of metabolic energy from As^{+3} oxidation. Under aerobic conditions, most of the arsenic in soils is in the As^{+5} form. Microbiological transformation of As^{+5} into As^{+3} is slow, and only up to 0.5% of arsenates are converted into arsenites. However, as a result of transformation, As^{+5} adsorbed on soil particles is released into the soil solution in the form of As^{+3}, which is much more mobile and toxic than As^{+5}. Therefore, microorganisms can increase the mobility of arsenic in the soil, thereby contributing to phytoremediation. Some *Bacillus* and *Shewanella*, strains were isolated from the As-contaminated sites, cultivated and characterized. These strains are the basis for the further phytoremediation, especially *Bacillus* spp., as they as rhizosphere microorganisms may bioabsorb arsenic from contaminated soils and by that to promote and facilitate soil phytoremediation.

Acknowledgements

This work was supported by grant # CARYS-19-179 from Shota Rustaveli National Science Foundation (SRNSF).

Author details

Gia Khatisashvili[1], Tamar Varazi[1*], Maritsa Kurashvili[1], Marina Pruidze[1], Evgeni Bunin[1], Kakha Didebulidze[1], Tinatin Butkhuzi[1], Elina Bakradze[2], Nino Asatiani[3], Tamar Kartvelishvili[3] and Nelly Sapojnikova[3]

1 Durmishidze Institute of Biochemistry and Biotechnology, Agricultural University of Georgia, Kakha Bendukidze University Campus, Tbilisi, Georgia

2 Ministry of Environment Protection and Agriculture, National Environmental Agency, Tbilisi, Georgia

3 Andronikashvili Institute of Physics, I. Javakhishvili Tbilisi State University, Tbilisi, Georgia

*Address all correspondence to: t.varazi@agruni.edu.ge

References

[1] Smedley PL, Kinnburgh DG. A review of the source, behavior and distribution of arsenic in natural waters. Applied Geochemistry. 2002;17: 517-568.

[2] Duarte AAIS, Cardoso SJA, Alcada AJ. Emerging and innovative techniques for arsenic removal applied to a small water supply system. Sustainability. 2009; 1:1-9.

[3] Mandal BK, Suzuki KT. Arsenic round the world: a review. Talanta. 2002; 58:201-235.

[4] Rosen BR, Liu ZJ. Transport pathways for arsenic and selenium: a minireview. Environment International Journal. 2009; 35:512-515. DOI:10.1016/j.envint.2008.07.023

[5] Cressey D. 'Arsenic-life' bacterium prefers phosphorus after all. Nature. 2012. DOI:10.1038/nature.2012.11520

[6] Butt AS, Rehman A. Isolation of arsenite-oxidizing bacteria from industrial effluents and their potential use in wastewater treatment. World Journal of Microbiology and Biotechnology. 2011; 27: 2435-2441. DOI:10.1007/s11274-011-0716-4

[7] Huang A, Teplitski M, Rathinasabapathi B, Ma L. Characterization of arsenic-resistant bacteria from the rhizosphere of arsenic hyperaccumulator *Pteris vittata*, Canadian Journal of Microbiology. 2010; 56: 236-246.

[8] Bhattacharya P, Jacks G, Frisbie SH, Smith E, Naidu R, Sarkar B. Arsenic in the environment: a global perspective. In Sarkar B, editor. Heavy Metals in the Environment. New York: Marcel Dekker; 2002. p. 147-215.

[9] Williams LE, Barnett MO, Kramer TA, Melville JG. Adsorption and transport of arsenic(V) in experimental subsurface systems. Journal Environmental Quality. 2003;32:841-850. DOI:10.2134/jeq2003.8410

[10] Zhang H, Selim HM. Reaction and transport of arsenic in soils: Equilibrium and kinetic modeling. Advances in Agronomy Journal. 2008; 98:45-115. DOI: 10.1016/S0065-2113(08)00202-2

[11] Satyapal GK, Ran S, Kumar M, Kumar N. Potential role of arsenic resistant bacteria in bioremediation: current status and future prospects, Journal of Microbial and Biochemical Technology. 2016; 8: 256-258. DOI: 10.4172/1948-5948.1000294

[12] Nies DH. Resistance to Cadmium, Cobalt, Zinc and Nickel in microbes. Plasmid.1992;27: 17-28. DOI:10.1016/0147-619x(92)90003-s

[13] Ahmann D, Roberts AL, Krumholz LR, Morel FMM. Microbe grows by reducing arsenic. Nature. 1994; 371:750. DOI: 10.1038/371750.a0

[14] Wolfe-Simon F, Switzer Blum J, Kulp TR, Gordon GW, Hoeft SE, Pett-Ridge J, Stolz JF, Webb SM, Weber PK, Davies PC, Anbar AD, Oremland RS. A bacterium that can grow by using arsenic instead of phosphorus. Science. 2011;332:1163-11666. DOI: 10.1126/science.1197258.

[15] Oremland RS, Newman DK, Kail BW, Stolz JF. Bacterial respiration of arsenate and its significance in the environment. In: Frankenberger WT, editor. Environmental chemistry of arsenic. New York: Marcel Dekker CRC Press; 2001. p. 273-296.

[16] Dopson M, Lindstrom E, Hallberg KB. Chromosomally encoded arsenical resistance of the moderately Thermophilic Acidophile Acidithiobacilluscaldus. Extremophiles. 2001;5: 247-255.

[17] Niggemyer A, Spring S, Stackebrandt E, Rosenzweig RF. Isolation and characterization of a novel As (V)-reducing bacterium: implications for arsenic mobilization and the genus *Desulfitobacterium*. Applied and Environmental Microbiology. 2001;67:5568-558. DOI: 10.1128/AEM.67.12.5568-5580.2001

[18] Anderson CR, Cook GM. Isolation and characterization of arsenate-reducing bacteria from arsenic contaminated sites in New Zealand. Current Microbiology. 2004;48: 341-347. DOI: 10. 1007/s00284-003-4205-3

[19] Silver S, Phung LT. Genes and Enzymes Involved in Bacterial Oxidation and Reduction of Inorganic Arsenic. Applied and Environmental Microbiology. 2005;71:599-608. DOI: 10.1128/AEM.71.2.599-608.2005

[20] Mukhpodhyay R, Rosen BP, Phung LT, Silver S. Microbial arsenic: from geocycles to genes and enzymes. FEMS Microbiology Reviews. 2002;26:311-325. DOI: 10. 1111/j.1574-6976.2002.tb00617.x

[21] Loy A, Lehner A, Lee N, Adamczyk J, Meier H, Ernst J, Schleifer K-H, Wagner M. Oligonucleotide microarray for 16S rRNA gene-based detection of all recognized lineages of sulfate-reducing prokaryotes in the Environment. Applied and Environmental Microbiology. 2002;68:5064-5081. DOI: 10.1128/AEM.68.10.5064-5081.2002

[22] Al-Humam AA, Zinkevich V, Sapojnikova N, Kartvelishvili T, Asatiani N. Biochips and rapid methods for detecting organisms involved in microbially influenced corrosion (MIC) [patent]. USA patent 15/949,400; 2018.

[23] Schraft H, Griffiths MW. Specific oligonucleotide primers for detection of lecithinase-positive Bacillus spp. by PCR. Applied and Environmental Microbiology. 1995;61:98-102.

[24] Qin X, Emerson J, Stapp J, Stapp L, Abe P, Burns JL. Use of Real-Time PCR with multiple targets to identify *Pseudomonas aeruginosa* and other nonfermenting gram-negative Bacilli from patients with cystic fibrosis. Journal of Clinical Microbiology. 2003;41:4312-4317. DOI: 10.1128/JCM.41.9.4312-4317.2003

[25] Zinkevich V, Beech IB. Screening of sulfate-reducing bacteria in colonoscopy samples from healthy and colitic human gut mucosa. FEMS Microbiology Ecology. 2000;34:147-155.

[26] Bunin E, Khatisashvili G, Varazi T, Kartvelishvili T, Asatiani N, Sapojnikova N. Study of arsenic-contaminated soil bacterial community using biochip technology. Water, Air and Soil Pollution. 2020;231:198.

[27] Islam F, Gault A, Boothman C, Polya DA, Charnock JM, Chafferjee D, Lloyd LR. Role of metal-reducing bacteria in arsenic release from Bengal delta sediments. Nature. 2004;430:68-71.

[28] Serrano J, Leiva E. Removal of arsenic using acid/metal-tolerant sulfate reducing bacteria: a new approach for bioremediation of high-arsenic acid mine waters. Water. 2017;9:994. DOI:10.3390/w9120994

[29] Chen C, Li L, Huang K, Zhang J, Xie W-Y, Lu Y, Dong X, Zhao F-J. Sulfate-reducing bacteria and methanogens are involved in arsenic methylation and demethylation in paddy soils. The ISME Journal. 2019;13:2523-2535.

[30] Hubert C, Nemati M, Jenneman G, Voordouw G. Corrosion risk associated with microbial souring control using nitrate and nitrite. Applied Microbiology and Biotechnology. 2005;68:272-282.

[31] Shavliashvili L, Bakradze E, Arabidze M, Kuchava G. Arsenic pollution study of the rivers and soils in some of the regions of Georgia. International Journal of Current Research. 2017;9:47002-47008.

[32] Stainier RT, Palleronia NJ, Duodoroff M. The aerobic Pseudomonas: A taxonomic study. Journal of General Microbiology. 1966;43:159-271.

[33] Saltikov CW, Wildman RA Jr, Newman DK. Expression Dynamics of Arsenic Respiration and Detoxification in *Shewanella sp*. Strain ANA-3. Journal of Bacteriology. 2005;187:7390-7396.

[34] Turpeinen R, Kairesalo T, Häggblom MM. Microbial community structure and activity in arsenic, chromium- and copper-contaminated soils. FEMS Microbiology Ecology. 2004;47:39-50.

[35] Bahar MM, Megharaj M, Naidu R. Arsenic bioremediation potential of a new arsenite-oxidizing bacterium *Stenotrophomonas* sp. MM-7 isolated from soil. Biodegradation. 2012;23:803-812. DOI: 10.1007/s10532-012-9567-4.

Chapter 5

A Call to Action: Incentivizing Arsenic Remediation

Bartlomiej K. Bancewicz

Abstract

Arsenic is a threat to human health. Long-term Arsenic exposure can lead to numerous cancers and non-carcinogenic diseases. Over 230 million across 107 countries are drinking groundwater Arsenic concentrations above the maximum concentration limit of 10 μg/L. The number of affected individuals is expected to rise in parallel with a growing dependence on groundwater, driven by diminishing surface water quality and quantity. A growing number of people will come in contact with Arsenic-contaminated water at new locations, while excessive pumping, geogenic processes, and industrial sources raise Arsenic concentrations at active groundwater sites. It is time to begin implementing Arsenic remediation techniques to save human lives, boost the global economy, and instill the foundations of a global collaborative framework. The continued research and development of remediation technologies is crucial, but these technologies will remain ineffective unless implemented. This chapter reviews the ongoing Arsenic crisis and suggests a simplified plan of action for resolving this problem. This is a transcontinental endeavor, which must begin with world leaders identifying and engaging new stakeholders. This will require education and awareness campaigns to boost involvement of the public sector, private sector, and the general public.

Keywords: Arsenic remediation, human health, groundwater, contamination, developing nations, incentives, water strategy, global collaboration

1. Introduction

Globally, there is a growing dependence on groundwater for drinking water, agriculture, sanitation, energy, and industrial processes. The rise in groundwater use is driven by a loss of quantity and quality of surface water sources, such as lakes and rivers. Surface waters are replenished by precipitation, groundwater recharge, and runoff from other water bodies. Quantity is reducing because of increased evaporation rates caused by global warming, increased population demands, and inefficiencies within water infrastructure and agricultural irrigation. Quality is worsening due to industrial and agricultural pollutants, and higher salinity caused by increased evaporation rates (salt constituents remain as surface water evaporates). Surely, groundwater is a practical substitute where surface water is diminishing, but its excessive use brings several challenges.

First, groundwater does not replenish as quickly as surface water; it relies on the vertical seepage of surface waters through permeable ground. Therefore, long-term pumping will lead to a shortage of groundwater, and a further shortage of surface waters which rely on recharge from groundwater sources.

Second, a continued loss of groundwater can lead to land subsidence. This is when land begins sinking due to a loss of supporting water weight beneath it [1]. In the United States, groundwater depletion rates continue to rise and overstress aquifers in both arid and non-arid regions [1]. The U.S. Geological Survey (USGS) approximated 1000 km^3 of groundwater depletion between 1900 and 2008, with an average depletion rate of 9.2 km^3 per year [1]. But, between 2000 and 2008 alone, the average depletion approached 25 km^3 per year [1].

And third, excessive or new groundwater use raises the likelihood of accessing contaminated drinking water. Excessive pumping may cause saline groundwater to migrate inland and upward, resulting in contaminated water supply [1]. Geogenic and industrial groundwater contaminants can, likewise, migrate to current pumping sites, or be encountered when pumping from new sites already hosting such contaminants. In either case, whether the site is old or new, growing groundwater dependence raises the likelihood of human, animal, and plant exposure to contaminants.

Geogenic contaminants are ones which originate from rock materials, through weathering processes that lead to their deposition into an aqueous phase, via natural soil/rock-water interactions [2]. According to the Environmental Protection Agency (EPA), common geogenic contaminants include Iron, Manganese, Chlorides, Sulfates, radionuclides, Fluorides, and Arsenic [3]. It is acknowledged that Fluoride and Arsenic are the two most widespread geogenic contaminants, affecting millions of lives across the world [2]. The latter of these is the primary focus of this chapter.

Due to its wide distribution across Earth's crust, at an average concentration of 5 mg/kg [4] (existing in over 300 known minerals such as Arsenopyrite, Cobaltite, Enargite, Gersdorffite, Löllingite, Orpiment, Pyrite, Realgar, and Tennantite [5]), over 90% of the world's Arsenic pollution comes from geogenic sources [6]. The remaining near-10% of pollution is assumed to come from industrial processes or products: mining activities [4]; preserved wood products [7]; pharmaceutical, glass, microelectronic, and optical industries [7]; and historical use in pesticides and insecticides [7].

Arsenic enters groundwater through mechanisms of weathering, oxidative and reductive dissolution of As-bearing minerals, and competitive exchange of As by other compatible ions like Nitrate, Phosphate, and Bicarbonate [6]. These mechanisms are influenced by pH, presence of organic matter, microbial activity, water table and sediment saturation, groundwater flow direction, topography, and marine transgression [6].

The constant presence of environmental Arsenic is a threat to human health. It is amongst other dangerous contaminants and water-borne diseases which severely stunt the growth of developing nations. Addressing the Arsenic crisis is in our best interest. This is a water problem worthy of attention, since it is expected to worsen with growing global groundwater dependence. In focusing our attention on Arsenic, we can gain experience in water problem solving and global collaboration. The takeaways of this experience are guaranteed to help us resolve other complex problems, such as those addressed by the United Nations (UN) Sustainable Development Goals.

2. Arsenic is a threat to human health

Chronic Arsenic exposure causes a number of severe health issues, both carcinogenic and non-carcinogenic. Arsenic exposure can lead to skin cancer, lung cancer, bladder cancer, liver cancer, prostate cancer, leukemia, neurobehavioral abnormalities, diabetes, skin disorders, cardiovascular diseases, and pregnancy complications (e.g. fetal mortality and preterm birth) [8]. Arsenic contamination in water contributes to contaminated soils, agricultural products, and fish, which all further raise the probability of oral ingestion [8].

The majority of Arsenic toxicity in humans is due to inorganic compounds containing pentavalent As^{5+} and trivalent As^{3+} [8]. The As^{5+} form is capable of replacing Phosphate in a number of biochemical reactions, due to its similar structure and properties [9]. Meanwhile, the As^{3+} form is 2 to 10 times more toxic than As^{5+}, since it suppresses the activity of over 200 enzymes while bonded with thiol (—SH) groups, affecting the functions of numerous body organs [8].

Key epidemiological evidence regarding Arsenic's carcinogenicity has come from studies conducted in Taiwan, Bangladesh, Chile, and Argentina, where people regularly consume drinking water containing high Arsenic concentrations of 150 μg/L [8]. Scientists have identified unconsolidated sedimentary aquifers, located within younger orogenic belts, as the most highly concentrated sources of groundwater Arsenic [6]. The largest of these orogenic belts are along the entire western coasts of North and South America, and through the top of Africa, southern Europe, central Asia, and South Asia [6].

In February 2018, the World Health Organization (WHO) published a fact sheet on Arsenic [10]. It was estimated that 140 million people across 50 countries drank water with Arsenic concentrations that exceed the maximum concentration limit of 10 μg/L [10] (before 1993, the permissible limit was 50 μg/L [6]). In May 2021 however, a focus paper published to *Geoscience Frontiers* reintroduced that estimate as 230 million people across 107 countries [6].

Of course, the continuity of weathering and industrial process contributes to higher Arsenic content in soil and water. But, the drastic rise in estimated Arsenic-impacted people, from 2018 to 2021, is less likely due to pollution than it is groundwater usage. Our growing dependence on groundwater use, which hydrogeologists refer to as the "Silent Revolution" [11], will contribute to new cases of Arsenic exposure—through excessive pumping at current pumping sites, and Arsenic encounters at new sites.

Currently, Asia and Europe are the most impacted by Arsenic contamination, with 32 and 31 countries affected, respectively [6]. Next come Africa (20 countries), North America (11 countries), South America (9 countries), and Australia (4 countries) [6]. Of the 230 million people affected, 180 million live in Asian countries such as Bangladesh, Pakistan, India, China, Nepal, Vietnam, Thailand, Burma, and Cambodia [6]. This is in stark contrast to 2.1 million people affected across 25 states of the U.S.A. [12].

Withal, every country on Earth can benefit from swift government actions that trigger Arsenic monitoring and remediation.

3. Arsenic determination, monitoring, and remediation

We must develop strategies to avoid and alleviate the impacts of Arsenic groundwater contamination. In areas where drinking water contains elevated concentrations, the immediate course of action is to find a safer source of water [13]. If a cleaner source is unavailable, Arsenic removal is recommended [13].

3.1 Arsenic determination and monitoring techniques

Each time a new groundwater (or surface water) source is considered, it must be checked for the presence of contaminants or pathogens. The preceding step to groundwater monitoring and remediation is determination. Determining the concentrations of various Arsenic species will guide informed decisions on how to track Arsenic movement, and ultimately, which removal technique to employ.

Known Arsenic determination techniques include Inductively Coupled Plasma Mass Spectrometry (ICP-MS) and High-Performance Liquid Chromatography (HPLC) [14]. These techniques contribute to sampling data that tracks the concentrations and types of Arsenic species (Arsenate and Arsenite) found in water. Arsenate and Arsenite may also be found in soil, along with four soil-Arsenic species: Monomethylarsonic Acid, Dimethylarsinic Acid, Arsenobetaine, and Arsenocholine [14]. Soil species are important to consider because they can migrate into groundwater.

Groundwater monitoring is carried out by special instrumentation in monitoring wells. Example instruments include electronic steel tapes, pressure transducers, and automatic dataloggers, amongst others [15]. These instruments provide information on water level, conductivity, temperature, saltwater intrusion, and aquifer properties, making them a useful asset in testing for pollutants [16].

However, due to time and financial constraints, it can be difficult to perform consistent monitoring across large areas over time [17]. For this reason, Geographical Information Systems (GIS) software has proven highly useful for Arsenic concentration mapping, over large spatial and temporal scales [17].

Several GIS interpolation methods (Inverse Distance Weighted (IDW), Ordinary Kriging (OK), and Empirical Bayesian Kriging (EBK)) have demonstrated successful application towards groundwater monitoring data in the U.S.A. [17]. One local-scale case study (~900 m^2) indicated OK as the most accurate method for predicting Arsenic concentration fluctuation over a 9-month period [17]. Interpolation methods will become an increasingly fundamental part of pollutant monitoring, but will still require in-person sampling to ensure their calibration.

3.1.1 Significance of monitoring at the global scale

In parallel with contaminant monitoring, our awareness of groundwater withdrawal is exigent. Globally, we must track our water usage to remain informed of its availability.

In his book *Corporate Water Strategies*, Water Foundry CEO William Sarni points to the importance of water-footprinting. Water-footprinting is, essentially, the practice of water accounting [18].

Under the European Union (EU) Water Framework Directive (WFD), Spain was the first country to take advantage of water-footprinting, so it may develop a transparent and multidisciplinary framework for water-policy decisions [18]. Considering the world's growing dependence on groundwater, it can be understood why water-footprinting will become an imperative practice.

The following is a list of six potential water-footprinting tools to explore: Water Footprint Network (WFN), International Standards Organization (ISO) 14046, World Business Council for Sustainable Development (WBCSD) Global Water Tool, Global Environment Management Initiative (GEMI), Water Initiative (General Electric, Water Resources Institute and Goldman Sachs), and Corporate Water Gauge – Center for Sustainable Innovation [18].

Ideally, in the age of Big Data, the culmination of continued water research will grant superior water management capabilities. The growing availability of information and discoveries in fields of hydrogeology, climatology, astronomy, physics, biology, chemistry, and human activity will improve our analytical and predictive methodologies for multi-faceted problems. In the context of water, this will translate to pinpoint accuracy of water resource availability and quality, and contaminant fate and transport.

DOI: http://dx.doi.org/10.5772/intechopen.99376

3.2 Arsenic remediation techniques

The global Arsenic crisis can be solved by implementing a combination of existing remediation techniques, at both active and prospective groundwater sites.

To date, a vast number of Arsenic remediation techniques have become well-documented, and continue to grow more efficient. These include conventional techniques, as well as emerging innovations in nanotechnology, and fully sustainable approaches.

3.2.1 Conventional remediation techniques

Conventional techniques are those which have the longest history of field use. These include: (1) oxidation, (2) coagulation-flocculation, (3) adsorption, (4) ion exchange, and (5) membrane filtration [19].

(1) Oxidation is an often-necessary preceding step for removal. Oxidation converts the trivalent Arsenite (As^{3+}) form to the pentavalent Arsenate (As^{5+}) form [18], to promote its extraction from groundwater.

Arsenite extraction by means of adsorption and precipitation is difficult, since it is non-charged at pH levels below 9.2 [19]. Arsenate is favorable because of its propensity to adsorb onto solid surfaces, and co-precipitate with metallic cations [13]. Due to the nature of these two Arsenic species, many treatment techniques incorporate a two-step approach with an initial oxidation step, followed by removal [19].

Numerous oxidizing agents can be used: atmospheric Oxygen, Hypochlorite, Potassium Permanganate, UV radiation, and even bacteria (e.g. chemoautotrophic Arsenite-oxidizing bacteria) [19]. Atmospheric Oxygen is frequently used in developing countries, due to its low cost and abundance [20], but it is also very slow, often taking hours or weeks to complete [19].

(2) Coagulation-flocculation is one removal option following oxidation. Positively charged Alum or Ferric agents are incorporated to neutralize the opposing forces which separate colloidal particles [20] (i.e. Arsenic ions dispersed in groundwater). This leads to the formation of larger particles, or flocs, which bind together and clump up as agglomerates [20]. Throughout this process, Arsenic gradually takes the form of an insoluble solid that becomes available for precipitation [19].

Northern Chile has been removing groundwater Arsenic with this approach since 1970 [13]. Through continued research efforts, coagulation-flocculation has become highly effective. It is possible to reduce Arsenic concentrations from 400 μg/L to 10 μg/L, at a rate of 500 L/sec [13]. Still, a major drawback of this technique is the production of large sludge quantities with high Arsenic concentrations [20]. The sludge treatment that follows this is costly, but necessary so as to avoid secondary pollution [19].

(3) Adsorption is a physical process that utilizes solids as a medium for substance removal from liquids and gases [19]. Adsorption often utilizes a vessel or column filled with a packed bed of solid adsorption media, such as activated Carbon, Iron-based adsorbents, nanomaterials, and low-cost agricultural or industrial by-products [20].

As water passes through the bed of material, Arsenic and other impurities adsorb onto the surfaces of the adsorbent column [20]. Eventually, the column becomes saturated with water and impurities, requiring regeneration via exposure to 4% caustic soda (NaOH) [20]. The result of this is caustic waste water with high Arsenic concentration [20]. As with sludge from coagulation-flocculation, this creates a secondary problem, since waste water must be neutralized to prevent Arsenic waste generation [20].

(4) Ion exchange is a physical–chemical process which transfers ions between a solution phase and a solid resin phase [21]. This technique is commonly used for water softening and Nitrate removal in drinking water treatment systems [21].

The resin itself is usually an elastic three-dimensional hydrocarbon network, with a high quantity of ionizable groups electrostatically bound to the resin [21]. These groups are "exchanged" for ions bearing similar charge, but stronger selectivity for the resin [21]. Arsenite groups must be oxidized to Arsenate in order for this method to be effective [21].

(5) Membrane filtration technologies are designed to be selectively permeable: membrane structures permit some molecules to diffuse through, while blocking others [22]. Membrane filtration can be used to filter bacteria, salts, and heavy metals in water [22]. Membrane thickness, pore size and spacing, and material are key design parameters influencing molecular selectivity, diffusivity rate, and membrane durability.

Current reverse osmosis and nanofiltration membranes can operate at 40–400 psi (276–2,760 kPa), while rejecting 96–99% of both Arsenate and Arsenite species in water [22]. The ability to mitigate both Arsenate and Arsenite with high efficiency will prove useful for future point-of-use treatment.

3.2.2 Nanotechnology for arsenic remediation

In addition to conventional approaches, the 21st century has witnessed the emergence of nanotechnology-based remediation techniques. Nanotechnology is the branch of science dealing with materials at dimensions of approximately 1–100 nm [23].

Nano-sized materials (nanomaterials) are advantageous in water remediation applications—particularly adsorption for contaminant removal [24]. This is because nanomaterials offer high surface area, high number of active sites, porous structures, magnetic nature, and photocatalytic activity [25].

The predominant mechanism of Arsenic adsorption is via complexation and ligand exchange on nanomaterial surfaces [25]. As such, nanomaterials make for excellent media in fixed bed adsorption vessels [20]. Media such as Cupric Oxide nanoparticles, Iron Oxide-based nanoparticles, TiO_2 nanoparticles, ZrO_2 nanoparticles, and Zero-Valent Iron (ZVI) nanoparticles have been investigated for Arsenic removal from water [20]. A comprehensive review of nanomaterials for Arsenic water remediation is provided by [25].

Overall, the high adsorption properties of nanomaterials translate to accelerated remediation rates, lower costs, less material usage, and a lower generation of hazardous by-products [24]. The high efficiency of nanomaterials makes them ideal for expedited remediation efforts in regions most severely impacted by Arsenic remediation (i.e. India and Bangladesh) [24].

3.2.3 Sustainable solution highlight: phytoremediation

Sustainable remediation solutions are ones which are low-cost, utilize little material, produce no waste by-product, and generate zero emissions. For this reason, phytoremediation is a viable option for long-term Arsenic removal.

Phytoremediation is a simple, cost-effective, environmentally-friendly approach for removing Arsenic from aquatic environments [26]. Particularly in developing regions of the world, this is a great alternative where complex manufacturing is unavailable for the fabrication of other remediation technologies. Phytoremediation uses living plants such as ferns, shrubs, and herbs used to remove Arsenic from soil-water systems, or to render it less toxic [27].

A worthwhile remediation strategy for the future is to build more constructed wetlands (CWs) that incorporate phytoremediation. CWs are artificial wetlands used for the treatment of contaminated surface and sub-surface water [26]. A 2019 study, conducted by [28], tested this strategy with a 355-hectare (3.55 km^2) CW in France [28]. Results showed that Arsenic was mitigated at the CW outlet, with an approximate efficiency of 23% [26]. The efficiency of contaminant removal in similarly artificial ecosystems is still poorly-documented [28], so their potential has yet to be fully realized.

There is still much to learn in the field of phytoremediation, such as the application of different species, transgenic plants, and antistress responses [26]. In one case study, soil samples from Thailand were used to explore the effectivity of 36 plant species [27]. The study identified two species of fern (*Pteris vittata* and *Pityrogramma calomelanos*), one herb species (*Mimosa pudica),* and one shrub species *(Melastoma malabrathricum*) as some of the most efficient plants for phytoremediation [27]. Amongst these plants, the ferns most proficiently accumulated Arsenic from soil, attaining concentrations of up to 8350 mg/kg (dry soil mass) [27].

On their own, phytoremediation plant species are a promising means of Arsenic removal. But, one downside to phytoremediation is that it takes longer than other remediation techniques. Most heavy metal accumulating plants take about five years to uptake significant levels of contamination, since they grow slowly and their roots only penetrate the topsoil (5–20 cm) [29]. Fortunately, a worthy solution to this problem lies in mycorrhizal networks.

3.2.4 Mycorrhizae to accelerate phytoremediation

Mycorrhizae are symbiotic relationships between plant and fungi species [30]. Fungi colonize the root system of plants, providing both increased water and nutrient absorption capabilities [30]. Meanwhile, plants provide fungi with carbohydrates formed during photosynthesis [30].

A study by [29] demonstrated mycorrhizal networks at the roots of two accumulator plants, *Pteris vittate* and *Cynodon dactylon*, through the addition of *Glomus mosseae* and *Glomus intraradices* fungi [29]. Both the *Pteris vittate* and *Cynodon dactylon* experienced significantly higher biomass and Translocation Factor (TF) for Arsenic uptake from soil.

TF is a metric used to quantify Arsenic (or other metal) uptake through the plant; TF is determined through the following ratio: $[\text{Metal}]_{\text{aboveground tissue}}/[\text{Metal}]_{\text{belowground tissue}}$ [29], where [Metal] is the mg/kg concentration of the metal species, aboveground tissue refers to stems and leaves of the plant above ground, and belowground tissue refers to plant roots closest to the metal species.

To further progress the phytoremediation approach, the author of this chapter recommends continued studies in the area of mycorrhizal inoculants. Indeed, fungal species introduce a tremendous improvement to an already impressive, sustainable approach. In continuing this area of research, great advancement potential lies with fungi species from the company Dr. Janerette's Eco-Friendly Fungi.

Since April 2021, the author has had the privilege of interning under co-founder and CEO of this company, Dozie Mbonu. Both Dozie and co-founder Dr. Carol A. Janerette, a world-renowned botanist and environmental expert, have extensive knowledge in mycorrhizae. Dr. Janerette's Eco-Friendly Fungi can provide fungi which guarantee supreme nutrient uptake, water absorption, root health and longevity, and tolerance to high soil temperatures, toxic heavy metals, extreme pH levels, and transplant shock [31].

To date, the company has achieved superior results in the categories of crop yields and multi-harvests. One reason for this is that plants with Dr. J's Fungi can grow exponentially longer roots, and wider aboveground biomass, on a scale unlike any other. Crop growth has been attained even in infertile grounds, such as the abandoned anthracite coal mine in Plymouth Township, Luzerne County, Pennsylvania, U.S.A.

Partnerships with Dr. Janerette's Eco-Friendly Fungi would guarantee advancements in phytoremediation technology, in addition to reforestation efforts, and an eradication of global food scarcity.

4. Incentivizing global arsenic remediation

To mobilize the partnerships, funding, and curated efforts of global Arsenic remediation, the author offers three primary incentives: (1) eliminate human health threats by progressing clean water infrastructure, (2) boost the global economy, and (3) instill a framework for future global-scale collaborative efforts.

4.1 Incentive #1: eliminate human health threats

In addressing the Arsenic crisis, we can significantly reduce human health exposure to Arsenic, similar contaminants, and even pathogens. If we plan accordingly, the same remediation technologies for Arsenic may have overlapping purpose in the removal of Fluorides, Iron, Manganese, Chlorides, Sulfates, or radionuclides. At the same time, we can expand our database of groundwater quality and quantity in the regions we act. Compiling such a database across 107 countries would be highly beneficial towards our future water management objectives.

While working in Arsenic-impacted regions, we can remain proactive on the challenges of water-borne illnesses and water scarcity. These challenges are a more common topic of discussion than Arsenic, since they affect a larger percentage of the global population.

The WHO reported, in June 2019, an estimated 785 million people lacking complete access to basic drinking water services [32]. Furthermore, at least 2 billion people use a feces-contaminated drinking water source—this contributes to the transmission of diseases such as diarrhea, cholera, dysentery, typhoid, and polio [32]. It is estimated that these diseases cause 485,000 diarrheal deaths each year [32].

Without proper drinking water and wastewater infrastructure, it becomes impossible to escape disease. This is a dismal truth, but, fortunately, it has been the focus of many UN agencies and initiatives, along with various government and non-government organizations across the globe. There is no denying how invaluable our continued Water, Sanitation, and Hygiene (WASH) efforts are. However, it is equally important to simultaneously address contaminants of growing concern like Arsenic. Groundwater dependence will lead to an emergence of Arsenic and similar contaminants, in parallel with newly established WASH services.

4.2 Incentive #2: boost the global economy

Water is one of the greatest determining factors for a singular nation's prosperity. Thus far, water's availability has dictated which countries are industrialized or still developing.

As stated by current World Bank Group President David Malpass: "Clean water is a key factor for economic growth. Deteriorating water quality is stalling economic growth, worsening health conditions, reducing food production, and exacerbating poverty in many countries" [33]. The August 2019 World Bank Report found that a lack of clean water access limits economic growth by one-third [33].

When water comes from cleaner, more accessible sources, people spend less time and effort to physically collect it, freeing the opportunity to be productive in other ways [32]. For example, women and girls in Sub-Saharan Africa walk an average of 6 kilometers (3.7 miles) each day to gather clean water [34]. Between all women and girls of Sub-Saharan Africa, an estimated 200 million hours is spent collecting water every single day—this is over 22,800 years of daily activity lost to collecting water which is not guaranteed to be clean [34]. Undoubtedly, clean water sources closer to home would empower women to explore income-generating opportunities [35], contributing to a country's economic growth.

Better water sources also translate to lower health expenditures, since people can remain more economically productive with a lower likelihood of falling ill, and lower incurring medical costs [32]. In the context of children, access to improved water would guarantee better long-term health and, accordingly, better school attendance [32], translating to an increased output of societal members pursuing personal dreams and ambitions. The direct result of this is a more productive society which, over the course of several generations, would produce a more productive nation.

Truly, the importance of helping developing nations is too often underestimated. According to Jim Yong Kim, Former President of the World Bank Group: "global development extends far beyond charity and has a greater impact on the global economy than most people think" [36].

Following the 2008–2009 financial crisis, strong economic growth in developing nations became an "engine" for the global economy, contributing to about 50% of all global growth [36]. Additionally, a significant portion of U.S.A. exports go to emerging markets and developing economies [36].

Considering the fact that the U.S.A.'s trade partners (e.g. China, France, Japan, Mexico, Canada, Germany, and the United Kingdom) also profit off of exports to developing nations, it becomes heavily apparent that the prosperity of developing nations is crucial to the global economy. The U.S.A. relies on its trade partners for economic growth, and vice versa.

When countries participate in the exchange of goods, economic growth ensues. Thus, it is in our best interest to ensure global water security and clean water resources.

Arsenic is a growing threat to human health, and its removal would have an uplifting economic effect on both developing and industrialized nations. Water is the ultimate determinant of economy.

4.3 Incentive #3: develop a global collaborative framework

Resolving the Arsenic crisis comes easy once we overcome its preceding obstacle: convincing one another to collaborate.

Unquestionably, this becomes a largely philosophical puzzle. Despite being of the common *Homo sapiens* species, we are largely divided on our ideas of social identity and morality, amongst many others. These divisions are influenced by the space–time perspectives and experiences of every single individual that has ever existed on Earth. Our values overlap in many ways, and conflict in just as many. Yet, amongst our many differences, water is an unavoidable commonality.

If there is any cause worth uniting over, it is water. Water keeps us alive. Without water, there is no life. Humans, animals, plants, bacteria; nothing in this world can survive without water.

Most of this world is privileged to be born into clean water access, but a staggering percentage is not. The battle for clean water is being fought heavily, by a great number of organizations and communities across the world, but it is still not enough. Often, these valiant efforts face great adversity through a lack of resources (material, financial, and/or labor) to install clean water services. The solution to this is the influx of support from worldwide public and private sectors, water and non-water industries, and the general public.

The power of such a joint effort has yet to ever be realized. Now, more than ever, we have the collective resources to spread information and implement technologies at an international level. Since the beginning of time, planet Earth has had all resources necessary to resolving the world water crisis. It is a mere matter of gathering said resources, and implementing them where needed. As a united species, made up of diverse individuals and experiences, we can figure this out.

In taking on a challenge as humanity, rather than as individual countries and affiliated social groups therein, we will gain insurmountable experience in global collaboration. We stand to gain immense efficiency as a species.

Designating and proclaiming a challenge to overcome (e.g. the global Arsenic crisis), and taking a truly committed approach to resolve it, would provide novel takeaways on the flow pathways of material resources, financial resources, and labor, within interconnected dimensions.

Our new experiences and knowledge on how we worked together, and the routes we took for communicating and distributing resources, can be applied towards all 17 of the UN Sustainable Development Goals. These goals address water sanitation and hygiene, world hunger and poverty, climate action, ecological health, clean energy, sustainable cities and infrastructure, gender equity, health and well-being, quality education, and the formation of global partnerships. These goals define and outline the current biggest challenges we are to face as humanity. Establishing partnerships now would serve us well in the mission to resolve all 17 of these Sustainable Development Goals.

As for another 200, 300, 4,000, or 10,000 years into the future, there is no telling what challenges will come to light. What is certain, however, is that global collaboration will lead us to our solutions.

5. Plan of action

The chapter author has created a simplified problem-solving approach that can be applied towards complex problems. Though this model will evolve, the presented format leaves appropriate space for the overarching considerations of a multi-faceted endeavor. This model is a four-step process, which follows:

1. Identify the Problem. Identify (A) why it is a problem, (B) what caused it, and (C) how to address it.

2. Brainstorm Solutions. For each potential solution, determine (A) the approach via (i) resource (materials, financial, labor) acquisition, (ii) design and manifestation, (iii) installation and maintenance; (B) the external parties which will become involved, their intended role, how to recruit them, and how to keep them engaged; and (C) the impacts of each solution on (i) progress towards solving the ultimate objective, on (ii) those that will participate (internal and external parties) in solving the problem, and on (iii) those that will be liberated of the problem.

3. Coordinate efforts and spread awareness. Ensure all parties agree on the brainstormed solution(s) and how to implement them. Decide how information and data will flow between parties, while maintaining constant transparency. Furthermore, keep the general public informed on all news in progress—good or bad.

4. Take action, take notes, and spread further awareness. Employ solutions and make note of their impacts. Repeat current solutions and refine as needed. Continue brainstorming new solutions to adapt to the evolving situation. Spread awareness of each action taken to gather further support, for both current and upcoming solutions.

This model can be repurposed to fit the context of solving the global Arsenic crisis, as shown below. Note: at this time the model does not reflect the details to steps such as partnership formation or finance acquisition. Instead, the model is intended to provide an overview of what to expect when addressing the Arsenic problem.

1. The number of people exposed to Arsenic-contaminated groundwater is rising.

   A. Exposure to Arsenic at elevated concentrations (above 10 μg/L) leads to a number of carcinogenic and non-carcinogenic health effects. This is causing illness and deaths across 107 countries, blocking potential social and economic growth worldwide. Addressing the Arsenic crisis will be a leap in the effort to provide universal access to water and sanitation services.

   B. The quantity and quality of many surface water sources is diminishing, in parallel with growing population demands and changing climate, leading to an increased global dependence on groundwater. This raises the likelihood of human exposure to Arsenic at elevated concentrations, either by excessive pumping at current sites, or by encounter of Arsenic at new sites. Meanwhile, ongoing geogenic and industrial processes contribute to the number of dissolved Arsenic species in groundwater.

   C. In the short-term, Arsenic must be removed from areas where alternative sources of clean water are unavailable. Removal efforts will be aided by the continued collection of groundwater monitoring data. Also, stricter government controls must be implemented to prevent further Arsenic pollution from industrial processes.

   For the long-term: phytoremediation and artificial ecosystems will mitigate Arsenic in soil-water environments; collected groundwater and contaminant data will aid in the tracking of Arsenic migration; and global warming must be decelerated and reversed to prevent the rising evaporation rates which endanger surface water quantity and quality.

2. Proposed solution: Utilize groundwater monitoring and remediation techniques to track and remove Arsenic.

   A. (i) Materials for each technology can continue to come from current suppliers to research and industry institutions. Financial resources will be attained through continued identification and engagement of stakeholders, reallocation of public and private sector funding, and donations from the general public. Increased labor can come from the general public, employing individuals to work in jobs for the manufacture and operation of each technology, as well as in field maintenance jobs at the point-of-use.

   (ii) For each technology, design parameters will continue to evolve at the research level, through meticulous lab and field studies. Continued design evolution will lead to reduced material and financial costs, less waste generation, and lower emissions during fabrication. The fabrication process for these technologies will be determined by the parties which design them.

(iii) Install remediation technologies in all areas where people are affected by elevated Arsenic concentrations. Maintenance operations will be taught to the communities where remediation is implemented; this will require a transitional period where stakeholders stay in these regions and educate communities until they are fully independent.

B. The external parties for this solution include any individual or group who is not already contributing, in some capacity, towards the efforts of universal water and sanitation services. The intended role of external parties will be decided by the organizations which recruit them.

Spreading awareness of this problem and creating financial incentives such as green bonds and prize competitions may entice new research and industry institutions, and the general public, to contribute. Once new contributors are involved, maintaining their consistent engagement will become a challenge. It is crucial to provide real-time data on a local and global level so all parties observe the progressing impacts of their efforts.

C. (i) The end-result of this solution is the elimination of Arsenic health risks to humans. Concurrently, the Arsenic remediation technologies may remove other contaminants, such as Fluoride, from the same sites. The remediated regions will experience boosts to their economic prosperity, precipitating economic growth worldwide. Ultimately, substantial effort will have been made towards expanding clean water access for humanity.

(ii) Both the internal and external parties will benefit from this endeavor. Research and industrial institutions will have received new grants which accelerate field studies and breakthroughs in current technologies. Meanwhile, all involved parties will have gained valuable collaborative experience, with takeaways on financial flows and camaraderie that can be revisited and/or revised for future endeavors at the global scale.

(iii) Those liberated of Arsenic exposure will experience gains in health, water infrastructure, and economy. The impacts of remediation will be more significant in developing nations, where Arsenic poisoning is most severe. Here, a greater number of people will be brought to good health—both the current and future generations of each family will benefit.

3. Mobilization and partnership formation.

Before the solution is embarked on, it must be fully realized and communicated amongst *all* members of the parties which intend to commence it. It is recommended that this begins with the United Nations, the World Health Organization, European Union, federal governments, and foreign-equivalents of the Environmental Protection Agency and Center for Disease Control. These are the parties which can influence transcontinental action.

Once these initial parties agree on the solution, they must expand their spheres of influence. This will be done through partnerships with government and non-government organizations (NGOs), and not limited to experts in the agriculture, utilities, education, information, healthcare, arts & entertainment, marketing, financial, construction, manufacturing, engineering, and policy sectors.

Such partnerships will make it possible to spread education and awareness of the Arsenic problem, so there becomes a widespread recognition of the problem and a higher willingness to contribute material, financial, and labor resources.

4. Implement remediation technologies and keep the world informed.

   The public sector, private sector, and general public will become emotionally invested in the Arsenic problem as they learn of it and follow its progress. Publicizing it on mainstream and social media outlets will be an effective means of raising awareness. Albeit, there is always the danger of public division when such issues become publicized, even when an extensive network of factual evidence exists (i.e. climate change). Nevertheless, this need not impede us from embarking on necessary action. Progress must be publicized truthfully and frequently, regardless of how it is received.

The author's goal is to bring this model, and its future iterations, to the attention of world leaders and water professionals. To that end, the chapter author is a stakeholder, committed to driving change by catalyzing action.

6. Conclusions

This chapter was written to address the growing concern of human health exposure to Arsenic in elevated concentrations. The following are the key takeaways of this review:

- Arsenic is a threat to human health. Above the concentration of 10 μg/L in water, chronic Arsenic exposure can lead to cancers of the skin, lung, bladder, liver, prostate, and blood, in addition to neurobehavioral abnormalities, diabetes, skin disorders, cardiovascular diseases, and pregnancy complications.

- The number of Arsenic-impacted people is expected to rise, in parallel with growing global groundwater dependence. Population demands, and diminishing quantity and quality of surface water is leading to a growing global dependence on groundwater. People are encountering Arsenic-contaminated groundwater at (a) new sites because they do not have enough information about where to collect clean water, and at (b) ongoing sites through excessive pumping that promotes the uptake of contaminant species. Also, the continuation of geogenic processes and industrial Arsenic release contribute to the concentrations of Arsenic in soil and water.

- Arsenic monitoring and determination will help track the spread and existence of contaminants in groundwater sources. GIS and water-footprinting will be recognized as necessary tools in the future of water.

- There are numerous Arsenic remediation techniques which can already be implemented to begin saving lives. The conventional techniques of oxidation, coagulation-flocculation, adsorption, ion exchange, and membrane filtration, as well as innovations in nanotechnologies and phytoremediation, are all worthy of worldwide implementation.

- Implementing existing remediation techniques will contribute to their further research and development, through field studies and data acquisition.

- Phytoremediation is the most sustainable approach. Phytoremediation is inexpensive and generates no waste nor emissions. Additionally, mycorrhizal networks can significantly boost the Translocation Factor and biomass of phytoremediation plants. Dr. Janerette's Eco-Friendly Fungi is a company which can provide fungal species for the advancement of lab and field studies of phytoremediation.

- There are three primary incentives for launching worldwide remediation efforts: (1) eliminating human health threats of Arsenic exposure and other contaminants, (2) boosting the global economy, and (3) developing a global collaborative framework.

Acknowledgements

I wish to express tremendous gratitude to IntechOpen for the opportunity to contribute to "Arsenic." Writing this chapter has been a rewarding experience on many levels; for my academic, professional, and personal growth. Sincere thanks to Author Service Manager Karmen Daleta for her assistance throughout the submission process. And lastly, thank you to all of the world's organizations actively contributing to a clean water future—I look forward to joining you in your efforts.

Conflict of interest

The author declares no conflict of interest.

Note from author

Helping humans live better, more fulfilling lives will lead to the progression of humanity and societies worldwide. The resulting benefits of such actions, alone, are incentive enough to pursue them.

Even if we stood to gain nothing; even if our collaborative efforts served absolutely no benefit to us, why let that stop us? Is it the fear of losing money? The fear of investing our time? For, if it is a matter of money or time, imagine it was your water at stake.

Water, above time, is the most valuable resource one can own.

What value does time hold, in the absence of water?

Time, with tainted water, is torture.

And, time with no water at all, does not exist.

Do not let go of this precious resource. Educate yourself on all that goes into its cleansing and conservation, so you may project your knowledge through a love that helps others, and yourself, gain greater value from your time.

Thank you for your time.

Author details

Bartlomiej K. Bancewicz
Lehigh University, Bethlehem, PA, United States of America

*Address all correspondence to: bbancewicz@gmail.com

References

[1] Groundwater decline and depletion [Internet]. Usgs.gov. [cited 2021 Jul 2]. Available from: https://www.usgs.gov/special-topic/water-science-school/science/groundwater-decline-and-depletion?qt-science_center_objects=0

[2] Ghosh NC. Geogenic contamination and technologies for safe drinking water supply. In: Water and Sanitation in the New Millennium. New Delhi: Springer India; 2017. p. 81-95.

[3] Epa.gov. [cited 2021 Jul 2]. Available from: https://www.epa.gov/sites/production/files/2015-08/documents/mgwc-gwc1.pdf

[4] Garelick H, Jones H, Dybowska A, Valsami-Jones E. Arsenic pollution sources. Rev Environ Contam Toxicol. 2008;197:17-60.

[5] Drahota P, Filippi M. Secondary arsenic minerals in the environment: a review. Environ Int. 2009;35(8):1243-55.

[6] Shaji E, Santosh M, Sarath KV, Prakash P, Deepchand V, Divya BV. Arsenic contamination of groundwater: A global synopsis with focus on the Indian Peninsula. Geosci front. 2021; 12(3):101079.

[7] Chung J-Y, Yu S-D, Hong Y-S. Environmental source of arsenic exposure. J Prev Med Public Health. 2014;47(5):253-7.

[8] Hong Y-S, Song K-H, Chung J-Y. Health effects of chronic arsenic exposure. J Prev Med Public Health. 2014;47(5):245-52.

[9] Hughes MF. Arsenic toxicity and potential mechanisms of action. Toxicol Lett. 2002;133(1):1-16.

[10] Arsenic [Internet]. Who.int. [cited 2021 Jul 2]. Available from: https://www.who.int/news-room/fact-sheets/detail/arsenic

[11] Researchgate.net. [cited 2021 Jul 2]. Available from: https://www.researchgate.net/publication/255592141_Intensive_Groundwater_Use_Silent_Revolution_and_Potential_Source_of_Social_Conflicts/link/54157b210cf2fa878ad3f4d2/download

[12] Arsenic and Drinking Water [Internet]. Usgs.gov. [cited 2021 Jul 2]. Available from: https://www.usgs.gov/mission-areas/water-resources/science/arsenic-and-drinking-water?qt-science_center_objects=0

[13] Quaglino S. Arsenic Removal Technologies. In: The Italian Water Industry. Cham: Springer International Publishing; 2018. p. 89-100.

[14] Pongratz R. Arsenic speciation in environmental samples of contaminated soil. Sci Total Environ. 1998;224(1-3):133-41.

[15] Groundwater monitoring [Internet]. Globalw.com. [cited 2021 Jul 2]. Available from: http://www.globalw.com/support/groundwater.html

[16] GlobalSpec. Groundwater Monitoring Equipment Information [Internet]. Globalspec.com. [cited 2021 Jul 9]. Available from: https://www.globalspec.com/learnmore/sensors_transducers_detectors/environmental_sensors/groundwater_monitoring_equipment

[17] Fischer A, Lee M-K, Ojeda AS, Rogers SR. GIS interpolation is key in assessing spatial and temporal bioremediation of groundwater arsenic contamination. J Environ Manage. 2021;280(111683):111683.

[18] Sarni W. Corporate water strategies. London, England: Routledge; 2018.

[19] Nicomel NR, Leus K, Folens K, Van Der Voort P, Du Laing G. Technologies for arsenic removal from water: Current status and future perspectives. Int J Environ Res Public Health. 2015;13(1): ijerph13010062.

[20] Bhardwaj A, Rajput R, Misra K. Status of arsenic remediation in India. In: Ahuja S, editor. Advances in Water Purification Techniques. Elsevier; 2019. p. 219-58.

[21] U.s E, OW, Office of Ground Water, Drinking Water. Ion exchange. [cited 2021 Jul 2]; Available from: https://cfpub.epa.gov/safewater/arsenic/arsenictradeshow/arsenic.cfm?action=Ion%20Exchange

[22] Unicef RJC. Safe water technology for arsenic removal [Internet]. Unu.edu. [cited 2021 Jul 2]. Available from: https://archive.unu.edu/env/Arsenic/Han.pdf

[23] What it is and how it works [Internet]. Nano.gov. [cited 2021 Jul 9]. Available from: https://www.nano.gov/nanotech-101/what

[24] Sayan Bhattacharya, Indranil Saha, Aniruddha Mukhopadhyay, Dhrubajyoti Chattopadhyay, Uday Chand Ghosh and Debashis Chatterjee. Role of nanotechnology in water treatment and purification: Potential applications and implications [Internet]. Researchgate.net. 2013 [cited 2021 Jul 9]. Available from: https://www.researchgate.net/publication/261994441_Role_of_nanotechnology_in_water_treatment_and_purification_Potential_applications_and_implications

[25] Siddiqui SI, Naushad M, Chaudhry SA. Promising prospects of nanomaterials for arsenic water remediation: A comprehensive review. Process Saf Environ Prot. 2019;126:60-97.

[26] Berg EC, Borges AC. Use of plants in the remediation of arsenic-contaminated waters. Water Environ Res [Internet]. 2020;(wer.1419). Available from: http://dx.doi.org/10.1002/wer.1419

[27] Visoottiviseth P, Francesconi K, Sridokchan W. The potential of Thai indigenous plant species for the phytoremediation of arsenic contaminated land. Environ Pollut. 2002;118(3):453-61.

[28] Lebrun JD, Ayrault S, Drouet A, Bordier L, Fechner LC, Uher E, et al. Ecodynamics and bioavailability of metal contaminants in a constructed wetland within an agricultural drained catchment. Ecol Eng. 2019;136:108-17.

[29] Leung HM, Leung AOW, Ye ZH, Cheung KC, Yung KKL. Mixed arbuscular mycorrhizal (AM) fungal application to improve growth and arsenic accumulation of *Pteris vittata* (As hyperaccumulator) grown in As-contaminated soil. Chemosphere. 2013;92(10):1367-74.

[30] NYBG.Org: Hidden partners: Mycorrhizal fungi and plants [Internet]. Nybg.org. [cited 2021 Jul 2]. Available from: https://sciweb.nybg.org/science2/hcol/mycorrhizae.asp.html

[31] Janerette's Eco-Friendly Fungi [Internet]. Janerettesfungi.com. [cited 2021 Jul 2]. Available from: https://www.janerettesfungi.com/cred

[32] Drinking-water [Internet]. Who.int. [cited 2021 Jul 2]. Available from: https://www.who.int/news-room/fact-sheets/detail/drinking-water

[33] Worsening water quality reducing economic growth by a third in some countries: World Bank [Internet]. Worldbank.org. [cited 2021 Jul 2]. Available from: https://www.worldbank.org/en/news/press-release/2019/08/20/worsening-water-quality-reducing-economic-growth-by-a-third-in-some-countries

[34] Reid K. Walk for water: Your 6K vs. theirs [Internet]. Worldvision.org. 2020 [cited 2021 Jul 2]. Available from: https://www.worldvision.org/clean-water-news-stories/walk-water-6k

[35] Water & the economy [Internet]. Water.org. [cited 2021 Jul 2]. Available from: https://water.org/our-impact/water-crisis/economic-crisis/

[36] Why investing in poor countries helps all of us [Internet]. Worldbank.org. [cited 2021 Jul 2]. Available from: https://blogs.worldbank.org/voices/why-investing-poor-countries-helps-all-us